D1280177

BUSINESS AS
UNUSUAL

Anita Roddick

ANITA
RODDICK
BOOKS

Anita Roddick Books
An imprint of Anita Roddick Publications Ltd
93 East Street
Chichester
West Sussex
PO19 1HA UK

www.AnitaRoddick.com / www.TakeItPersonally.org

First published by Thorsons, an imprint of HarperCollinsPublishers 2000
This edition published by Anita Roddick Books 2005

All pictures © The Body Shop unless otherwise stated
Cover: Joel Anderson, photographer – www.joelanderson.com & design by Wheelhouse
Creative Ltd – www.wheelhousecreative.co.uk

A catalogue record for this book is available from the British Library

ISBN: 0 954 3959 5 6

Printed in the United Kingdom by Burlington Press Ltd
Printed on 100% recycled paper

Distributed in the United Kingdom by Airlift Book Company, 8 The Arena, Mollison Avenue,
Enfield, Middlesex, EN3 7NL

Distributed in North America by Chelsea Green Publishing, PO Box 428, 85 North Main
Street, Suite 120, White River Junction, VT 05001, USA

To my grandchildren, Maiya, Atticus and O'sha, in the knowledge that they'll end up like their parents and grandparents (and of course, *Mother Jones*) — ballsy, truth-telling, free-thinking, heart-bleeding, myth-debunking, non-conforming and hell-raising activists.

With thanks to all franchisees past and present and employees past and present who've helped shape The Body Shop into a skin and haircare company with attitude.

THE
BODY
SHOP

*out
standing
in
her
field*

**women in business are
cropping up all over**

CONTENTS

I like the title
'Business as Unusual'
for the book - But
how about the sub-
title "Management
by falling apart at
the seams"?

Because that's
what I feel likes at
the moment!
Best wishes Anita

Royal Mail DENNISCARDS & PUBLISHING

Mail Centre

1ST

Anita Roddick Books
93 East Street
Chichester
West Sussex
PO 19 1HA

© E.T.W. DENNIS & SONS LTD. SCARBOROUGH (01723) 500555

ALL BRITISH PRODUCTION

5 012491 000167

L016229L

My postcard to the publishers about this book

OH, THE JOY OF THE JOURNEY!

'You are not remembered for what you do in business, only for what you do in civil society.'

I have never been able to discover who said that – it could even have been me – but it neatly sums up my business philosophy, and when I first heard the phrase it was an epiphany for me. Of course I passionately believe that successful businesspeople should contribute to organizations that can make the world a better, safer, more civilised place. Not only because that is the right thing to do, but because only the truly myopic are unable to recognise that a fairer world is better for both the soul and the bottom line.

But it's more than that. I have never believed that business was in a separate compartment from civilising the world. That's why I have always been an activist, an agitator and an entrepreneur rather than a conventional business leader. On the other hand, I did manage to found and build up a company which is said to be one in the top 30 worldwide brands. It now has over 2,000 stores in 51 countries serving some eight million customers a day, so I must have done something right.

At first, it seemed simple. I just wanted to go in the opposite direction of the cosmetics industry. I wanted to stand for more than a mere bubble-bath. I refused to talk about our products as if they were the body and blood of Jesus Christ. But over the following quarter century, the company became enormous and enormously complicated too. And I wanted to go further – asking permission from customers, from employees, from suppliers and from the local community to move business from private greed to public good. They gave me permission, and despite all the predictions from business commentators, The Body Shop grew into one of the most successful retailers in the world.

But it was uncharted territory.

I then found myself on an extraordinary journey, a journey which made me even more passionate about the urgent need for a new model of business. It was also a journey through a minefield, with every explosion a reminder – as if I needed one – of how fragile the goals we'd set ourselves were.

That journey is the subject matter of this book.

Looking back from the vantage point of a new century, I can trace how the path I've taken has radicalized me. Whether it was a sally into the forests of Sarawak to photograph illegal logging or up the Amazon to set up the first direct trading link with the Kayapo, it gave me a glimpse of the way in which global corporations threaten to engulf life, not just in our society but also in the communities where businesspeople simply never go. The outrage I felt was no abstract emotion. It grew stronger the more I travelled, the more I met people forced to eke out a bare existence on the margins of the world, denied their fundamental human rights by the fickle voracity of the global business juggernaut.

I'd always known I was in uncharted territory, but sometimes the landscape I found myself in was so alien that the people I thought were friends were actually enemies – and vice versa. Talk about a recipe for paranoia. Life in a minefield isn't easy. There have been many dents in my optimism and challenges to my convictions – our libel action, our mistakes in the American market, that relationship with the Kayapo I mentioned – but this was ultimately a good thing. When you've had some years when everyone – including the stock market – says you can do no wrong, it's good to be reminded that you're running the same race as the rest of the world. What doesn't kill you makes you stronger.

Yet that sentiment is actually too harsh. 'Be kind,' Harvey Keitel writes on Kate Winslet's forehead in the movie Holy Smoke. That's what I wanted to do to the business world – nurture a revolution in kindness. But underlying that ambition has been the recurring theme of my role in the company I founded, now that it has grown into such a huge and complicated organization. So this book actually has another subject as well. It is the story of how I managed to maintain some intimate part of myself – the original core, if you like – in a business gone global. I have had to constantly reinvent the role of the founder-entrepreneur. That's tough when your natural tendency is towards a gleeful anarchy. There are no roadmaps, no instruction manuals. Passion is your guide. Instinct tells you where to go when a challenge arises.

So this book is more than a chronicle of 23 years at The Body Shop, more than a condensed manual for the wannabe business radical. It is also about one individual's attempt to marry the often impersonal wants of a successful business with the very personal needs of a successful businessperson. Despite the enormous constraints of a global company, despite the general intractability of life, I need to find new ways to push the limits of business, to change its language, to make it a force for positive change.

That's what I mean by business as unusual.

DON'T ACCEPT IT.

THE TRUTH IS THAT 'FREE TRADE' AS ITS VICTORIAN FOREFATHERS PROMOTE

IT WAS ABOUT THE FREEDOM OF COMMUNITIES TO TRADE EQUALLY WITH

EACH OTHER. IT WAS NEVER INTENDED TO BE WHAT IT IS TODAY – A LICENC

FOR THE BIG, THE POWERFUL AND THE RICH TO RIDE ROUGHSHOD OVE

THE SMALL, THE WEAK AND THE POOR ...

WHAT'S BUSINESS ALL ABOUT?

I suppose you might call it a defining moment. The year was 1987, The Body Shop was still in its protoplasmic period, but the Confederation of British Industry had chosen us as the Company of the Year, which meant I had to pick up the award.

There they sat, the captains of industry, the bankers, the analysts, the journalists, just about the entire British financial establishment, all lined up and holding their collective breath as I mounted the podium.

It was irresistible. Every provocative bone in my body was aching to act up. I had prepared what I hoped was a mildly inflammatory speech. When I showed it to Gordon, my husband, he advised me to go ahead and challenge them all. When I showed it to Jilly Forster, a close friend who also happened to be our PR person, she told me I had to say what I had written. As I started to speak, the tension was such that I could see tears streaming down her face. I could also see our brokers, sitting stony-faced on either side of her.

What did I say? I stood up for entrepreneurs against big corporations. I told the business titans they were stuck in the past – 'dinosaurs in pin-striped suits'. 'I've never met a captain of industry who made my blood sing,' I said, at which point I looked up to see Robert Maxwell, of all people, walking out. For all I know, he was just going to the Gents, but still I felt a surge of pride. Thanks to Captain Bob, my blood did sing after all.

That speech was a watershed for me. I had the optimistic sensation that The Body Shop way of doing business was the inescapable future.

In the decade or so that followed, the story got a lot more complicated – richer, but darker too. History did strange 180 degree contortions. The Berlin Wall came down (so did Robert Maxwell, for that matter). The maximalisation of the 1980s became the minimalisation of the 1990s. I became more controversial among the business elite: one captain of industry once went on the record to describe me as 'frenetic and self-righteous'. The world kept changing, but so did I.

The 'self-righteous' label probably said more about him than it did about me, but 'frenetic' had a touch of truth in it, because I am a firm believer that entrepreneurs have a nomadic soul. I have certainly always found it hard to keep still. It helps us understand the changing environment – for modern life as well as business – and it confronts us with the truth, as it has confronted me. Some journeys are more about business, but some are just for the truth, in countries we generally regard as rich as well as those we regard as poor.

Journeys have always provided me with insight, and two of them in particular. One was the journey to the furious meeting of the World Trade Organization (WTO) in Seattle at the end of 1999 which ended with extraordinary scenes on the streets as the world's anti-globalization activists, overwhelmingly peaceful, were confronted with the brutal determination of large numbers of extremely unpeaceful police. It was this journey that culminated in me being tear-gassed – a formative experience and something that few CEOs of major international companies can boast.

But it was another journey to the USA that opened my eyes most of all, travelling through the so-called 'Black Belt' of Louisiana, Mississippi, Alabama and Georgia for my first look at extreme poverty in the USA, the richest country of all. I found a guide in Jacob Holdt, a Danish 'vagabond' photographer who had spent 30 years roaming America photographing rural black communities. I was scheduled to give a talk in New Orleans and Jacob was making the journey to revisit people he had spent time with over the past 30 years. Together we visited shacks and prison communities in the forgotten underbelly of America.

When I first met Jacob, the first thing I wanted to do was hand him a bottle of our Brazil nut hair conditioner – his hair was as rough as straw and he had a long plaited beard which he rolled up when he went into cities. I quickly learned Jacob's personality was determinedly passive. If ever he was in adversarial situations, he would gently talk his way out of things. Once, when every inch of our truck was filled with wayfarers, itinerants and hitchhikers who would normally strike fear in any of us – Jacob never passed anyone on the highway – I studied how softly he spoke and how intently he listened. In our society, gentleness is often viewed as ridiculous or

I think racial prejudice is like a hair across your cheek – you can't see it in your behaviour, you can't find it, you just keep brushing against it.

insincere; Jacob showed me that nothing is more powerful or persuasive. Treating people with respect and kindness is how he has survived the hazards of the life he has spent gathering the stories of the marginalized and powerless.

What I saw whilst travelling with Jacob astounded me. It was my first encounter with real poverty in any Western country. During the trip, I also saw the overwhelming power of television. Its pale blue light flickered in the broken-down shacks 24 hours a day, pacifying the mind and perpetuating the myth that material wealth defines self-value and self-worth.

For the last 20 years I have experienced how wealth can

make you insensitive to the human condition and during that journey with Jacob I decided that I'd do anything and everything not to allow this insensitivity to happen to me. That journey provided me with yet another antidote to comfort and complacency. It helped illuminate the current state of human affairs.

This hopelessness and poverty in the midst of wealth is a stark reminder that the 'economic problem', as John Maynard Keynes put it, is not solved. We have had record booms in the US economy, we have had stock markets soaring over the 10,000-point barrier, but we have this looming human catastrophe as well. And poverty drives other crises. It drives desperate people to over-exploit their resources or allow them to be over-exploited. It drives them into drug dealing or terrorism. And it also drives them into the waiting arms of rich countries as military, economic or political collaborators who perpetuate the poverty. In fact we've institutionalized it.

I sometimes wonder why we're not more outraged by the fact that three billion people live on less than $2 a day while the wealthy have stashed away $8 trillion in tax havens. They certainly don't seem to be picking up the tab on world poverty. Despite the astonishing wealth of the so-called 'long boom', a fifth of the human race still does not even have access to proper food or clean water.

THE RISE OF THE NGOS

This is the reality of the modern world, but it has spawned the beginning of a much more hopeful trend, and it's one that absolutely thrills me. I knew the business environment was changing, but until my flight to Seattle I hadn't understood just how much this was due to the rise of the NGOs. All the focus groups, futurists and pollsters employed by the corporations failed to predict the NGOs' extraordinary influence, yet their emergence could still change everything. There are now at least 100,000 NGOs working on green issues alone all over the world. Some of them are persecuted and embattled, yet some are increasingly powerful.

One question from Greenpeace by fax to a food manufacturer in 1999 was enough for them to take **GM ingredients** out of baby food. Indonesian NGOs helped bring down the dictator President Suharto.

The role of NGOs is to be the beneficial aspect of globalization. Their vigilance around the world makes the great abuses which humanity once brushed aside visible for all to see. Together, they now represent billions of people who are determined that trade should be more equitable, fair and just – often the least powerful people, those whose voices are heard the least. The NGOs face the suspicion of the left because they are too close to the corporate world and the suspicion of the right because they threaten to change the way business is carried out. But, taken together with the terrifying rise in poverty, they represent an entirely new climate for business. All over the world we can see unusual alliances being formed between human rights groups and educational institutions, alternative trade associations and progressive consumer groups, and often they are in partnership with business too.

The trouble is that much of the corporate world still fails to recognize their significance, or what they represent, or the catastrophe of poverty. Open up a typical management book and you will find it hard to avoid words like 'leadership', 'team-building', 'culture' or 'customer service', but you'll be lucky if you find words like 'community', 'economic poverty', 'social justice', 'ethics', 'love', 'care' or even 'spirituality' – a word that is truly kept in the closet! Not to mention an MBA curriculum that fails to include subjects such as social and ethical accounting, human rights or gender perspectives in the workplace. 'Words enable worlds,' said the philosopher Ludwig Wittgenstein and the rise of the NGOs may well be pivotal in fashioning some new worlds. But there is a long way to go and in the meantime we have to face up to some of the following problems.

FOOTLOOSE BUSINESS

If you look at the way some businesses are behaving in many corners of the world – the places most business leaders never visit – you can see them alienating humanity in so many ways. I have seen, and still see, corporate crimes in abundance. Industry after industry seems perfectly happy to use sweatshops and the globe is quickly becoming a playground for those who can move capital and projects quickly from place to place. When business can roam from country to country with few restrictions in its search for the lowest wages, the loosest environmental regulations and the most docile and desperate workers, then the destruction of livelihoods, cultures and environments can be enormous.

From Europe to the US to Taiwan to Malaysia, each country is just another pit stop in the race to an ever-improved bottom line. The new frontier is Asia, where wages and environmental standards are still lower and human rights abuses are suppressed in an even more sordid way. **The new nomadic capital never sets down roots, never builds communities. It leaves behind toxic wastes, embittered workers and indigenous communities driven out of existence.** I spend much of every year travelling around the world, talking to the victims of globalization, people like small farmers in the US, scores of whom go out of business every week. Half a century ago there were a million black farmers in the US; now there are 1,800. Globalization means that subsidies go to the big farms, while the small family farms – the heart of so many small town communities – go to the wall. I have also visited dark cramped factories where people work for a pittance for 12 hours a day without a day off. They are vic-

'We are not allowed to talk to each other or go to the bathroom,' one Asian worker in a garment factory told me. This wasn't in Seoul or Sao Paulo: it was in San Francisco.

tims, not just of globalization, but of a new creed – also a very old one – that simple power, economic or military, can do as it likes.

ENTRENCHED GREED

One of the key problems of the business world is that greed has become culturally acceptable. New York Stock Exchange chair Dick Grasso demanded a compensation package that included a one-off deferred pension payment of $140 million. Tyco CEO Dennis Kozlowski used to spend $15,000 on hat-stands for his office.

Or take the example of Walt Disney, chosen by the US corporate watchdog magazine *Multinational Monitor* as one of the 10 worst corporations of 1996 because of its refusal to pay decent wages to contract workers in Thailand, Haiti and even in the US. Haitian contractors producing children's clothing under licence to Disney were paying workers 28 cents an hour, or about 7 cents for every garment they made, while in other factories producing Disney clothing, workers earned as little as $1 a day – 12 cents an hour. The workers' call for a living wage – 58 cents an hour – fell on deaf ears at Disney. To interfere would be 'an inappropriate use of our authority', said a Disney spokesperson. It would take one Haitian worker producing Disney clothes and dolls 166 years to earn as much as Disney president Michael Eisner then earnt in one day. Eisner isn't even one of the seven richest men in the world, whose pooled wealth would – it is said – be sufficient to eliminate global poverty.

There is a direct link between this kind of greed, once it becomes widespread in society, and some of the worst social shocks of the day – like the children who murder each other just to get their Nike trainers. Greed without legal and moral constraint can destroy everything worthwhile in life. Wealth can corrode humanity and alienate the wealthy from the human condition.

This is not just a theoretical issue for me. The Body Shop has been successful enough for me to be wealthy compared to the vast majority of the world – though we have resisted the kind of corporate greed I've just described. I know it would be all too easy for me to sink into the seductive comfort of just being a wealthy business-woman. It would be an easy life, but a life without fighting for anything has a hint of death about it, so I do what I can to keep striving, travelling and fighting. What I see around me also helps keep me alive and, as I have told my children, when I die every penny I've ever earned will be given away to human and civil rights activists.

CASINO CAPITAL

'When the capital development of a country becomes a by-product of the activities of a casino,' warned Keynes, 'the job is likely to be ill-done.' In our generation, the casino analogy has also been used by the development economist Susan George as a way of describing how the financial system can be mendacious to a majority of the world's population. It is a world economic casino overseen and policed by no one, with central bankers acting as croupiers. Crowded about the tables are players using commodities, currencies and derivatives as chips. Outside, politicians act as greeters, like retired prizefighters yelling, 'Great to see you!' and bidding us all inside.

But in this casino it is those who run the operation who reap the spoils. After two centuries or more of world economic growth, the benefits are still only felt by a minority. And the goods for sale in the global economy are usually produced not by the world's victors, but by its victims, not by a charming band of little people singing, 'Hi ho, hi ho, it's off to work we go,' but by desperately poor people, often children, engaged in forced and bonded labour, and struggling in sweatshops, not in the democracies promised by the political economists, but in brutal dictatorships. And the results are not the promised rising standards of living, but poisoned air, water and land, and increasingly unequal wealth.

BEGGAR MY NEIGHBOUR

In a global market, where multinational corporations compete on price and cost cutting, the cheapest places on Earth for natural resources or labour are precisely those nations which place no value on democracy, human rights or environmental protection.

Take Burma, for example, ruled by one of the most repressive and barbaric regimes in the world, one which has ignored free democratic elections and imprisoned the victor, the heroic Aung San Suu Kyi. Yet what is a human rights disaster for its citizens is regarded as an investment opportunity by the big energy corporations. When Unocal Corporation of California and the French oil company Total formed a joint venture with the military dictators of Burma to extract gas from under the sea, some 40 miles off the coast, the military destroyed entire villages, seized property and attacked, raped and tortured innocent villagers, many of whom had been forced to labour as slaves on the pipeline. Of course that's not what the oil companies wanted, but they need to understand the consequences of doing business with a dictatorship.

UNFREE TRADE

It isn't just oil, and it isn't just Burma. The principle of 'free' trade applies to every other product or commodity, no matter what the cost in terms of low wages, spoiled environment and crushed workers' rights. And there is always some place in the world a little worse off, a little riper for exploitation.

The 'free market' conjures up fuzzy notions of free and equal individuals exchanging hand-made shoes for homegrown vegetables in a village square. This bucolic picture carries with it a notion that a job is just an individual contract between equals – but it hasn't been like that. Colin Hines from the International Forum on Globalization reports that after NAFTA, around 2,000 factories moved from the USA to operate in the border region in Mexico, virtually unhindered by lax environmental and labour regulations. NAFTA is the North American Free Trade Agreement – or, as some campaigners put it, Not Another Fucking Trade Association – and it has relaxed regulations to a dangerous extent. The environmental repercussions of more than 300 companies opening up works on the Mexican border with Texas are appalling. As many as 400,000 people live there without sufficient housing, running water, sewerage, pavements or electricity. An open canal carries 55 million gallons of raw sewage for 17 miles alongside the Rio Grande, polluting drinking water wells and the river itself.

Free trade is one of the greatest deceptions. Ask yourself whether the market is really free, and free for whom or for what. And how much of the world economy is based on trade? Every day, computer screens on money markets around the world oversee a global flow of no less than $1.5 trillion. The terrifying thing about that figure is that only about 5 per cent has anything to do with trade at all, let alone 'free' trade between communities. What it has to do with is money. That means that 95 per cent of that enormous sum is speculation and froth. It's money making money.

BLIND WORLD GOVERNMENT

The problem is that we have a world trading system that is blind to this kind of injustice, presided over by the World Trade Organization. The WTO and other multinational bodies may be all that guards the world from the might-is-right demands of the current administration in the USA. But the trouble is that the unelected trade officials that run it also have, in their own way, a bias towards the most powerful, with the right to overturn local laws or safety regulations whenever they say they interfere with free trade.

As well as being world government by default, the world trading system is also government that is at least partially blind. It looks at the bottom line, but can't see anything else. It can recognize profit and loss, but it cannot recognize human rights, child labour or the need to keep the environment viable for future generations. It is government without a heart, and without a heart you find the creativity of the human spirit starts to dwindle too.

I believe those who are now in control – the economic governments, politicians and businesspeople – are capable of driving us over the edge. Global planning institutions like the WTO, World Bank and the IMF are ignoring mounting evidence of an impending social catastrophe that will leave widespread, dangerous inequality and insecurity. Whatever their intentions, these institutions are simply not working for the majority of humanity.

The danger is that the current trading system undermines anyone who tries to do what we are doing. **Businesses which forego profits to build communities, or which keep production local rather than employing semi-slaves in distant sweatshops, risk losing business to cheaper competitors without such commitments and being targets for takeover by the slash-and-burn corporate raiders.**

THE SOULLESS DESTRUCTION OF MONOCULTURE

I am often asked what I fear most about the way business is going. My biggest fear is seeing not just the planet's business, but also the planet, being controlled by a handful of gigantic transnational corporations. You can see the beginning of this in the way that global brands are starting to raise our children. They entertain them, feed them, clothe them, medicate them, addict them and define the ways in which they relate to each other. By the age of seven, the average American child will be seeing 20,000 advertisements per year on television. By the age of 12, that child will have an entry in the massive marketing databases used by companies.

This kind of global monoculture wreaks a soulless kind of destruction. Not just on families, but on family farms. It means the proliferation of one-crop nations in a continent like Africa, with the whole crop being produced for export. It means millions of people who are no longer able to look to their land to supply their basic food needs. The idea is that this kind of market economy will raise people to a level of consumer benefit – so much so that they will abandon their homes and villages and crowd into the cities in search of monetary wealth – but it just isn't working.

With fewer and fewer corporations controlling more and more of the world's trade, there is an ever greater need to know more about the practices of these large faceless organizations. Redress is very slow – there is still no compensation for the thousands of victims of Union Carbide in Bhopal, for example, or for the Ogoni in Nigeria, or for countless other voiceless and powerless victims trampled underfoot in the rush to accumulate ever-increasing profits. But consumers are beginning to resist and some businesspeople are looking for a better way.

Back in 1993, I was invited to give a talk to the International Chamber of Commerce at a meeting that was to be held in Cancun. I had just returned from two weeks being with peasant farm workers in Mexico and had also seen for myself some of the results of callous business practices in the tobacco fields. I have been in places around the world where capital has newly, and temporarily, alighted, and where I've held mutated babies genetically handicapped by toxic wastes dumped in local streams. But this example was particularly upsetting, for the babies were being born without genitalia. Scientists had tracked down the cause to the pesticides used on the nearby fields.

The American tobacco companies which bought the crops grown there refused to accept any responsibility because the fields did not belong to them. So, knowing that representatives of the same companies were going to be in the audience in Cancun and that this kind of corporate indifference was endemic among the people I would be speaking to, I gave a talk about the key issue of the moment: NAFTA. And I showed them the slides of my journey with the Huichol Indians in the Sierra Madre.

I don't quite know what response I expected. I thought there would be some reaction, even if they howled me off the stage. But there was absolutely none – no embarrassment, no sense of outrage, just a collegiate sense of good manners. I was chilled by the experience because these were no ordinary people. Leaders in world

What is the business of business? To create wealth? To inspirit the economy? To create jobs? To meet the needs of society? Yes. But there is more. The final goal of any human activity, and any business must show us how to be effective, is to create a world moral order – a world ethics network.

Peter Koestenbaum

business are the first true global citizens.

As businesspeople, we have world-wide capability and responsibility; our domains transcend national boundaries. Our decisions affect not just economies but societies, not just the direct concerns of business, but the world problems of poverty, environment and security.

If business comes with no moral sympathy or honourable code of behaviour, then God help us all.

I don't think anyone would argue that business now dominates the world's centre stage. It is faster, more creative, adaptable, efficient and wealthier than many governments. Microsoft could fund the National Health Service, the Royal Navy, the RAF and the Army for a year, all by itself, and still have change to spare. Only 27 countries have a turnover greater than the combined sales of Shell and Exxon. Half of everything spent by British consumers goes into the coffers of just 250 companies. So in terms of power and influence, you can forget the Church and forget politics, too. There is no more powerful institution in society than business. It is more important than ever before for business to assume a moral leadership in society.

The trouble is that many of our business leaders are not equipped to deal with this new responsibility. For one thing, they are locked into the narrow economic thinking of the past. Our modern age is unique in the way economics overrides everything of value. Businesses have become hypnotized by the bottom line and forgotten their moral obligations to civil society. The message I have been repeating in every speech I've made and every article I've written on business in the last 20 years is that we must include in our measures of success enough to sustain communities, cultures and families, and the consequences will be severe unless we do so. **The prevailing view of trade could be described as commerce without**

conscience. And conscience is a key to the way out. The conscience is partly that of business itself, partly their customers and the public, and partly a matter of simple regulation.

CHANGE BY REGULATION

Let's take regulation first, because business needs a strong framework to thrive. Without a legal framework and constraints, there will always be a tendency for it to become criminal. You will now find that's why the second largest industry in the world is now the illegal drugs industry and one of the biggest sectors of economic growth is child sex.

'Goods produced under conditions which do not meet a rudimentary standard of decency should be regarded as contraband and ought not to be allowed to pollute channels of interstate trade.' Roosevelt's quote is more valid than ever today. And as globalization ties the world together more tightly, activities that are unaccountable and illegal are paradoxically increasing.

We need business to be accountable and to base its international behaviour on the charters and treaties on sustainability and human rights so happily signed by governments the world over, but so easily ignored by them. Regulation has to penalize businesses when they break the rules.

CHANGE BY CONSUMER CONSCIENCE

If regulation is not likely to be implemented, the conscience of ordinary people becomes of pivotal importance. And that's where the NGOs come in, monitoring the activities of the big corporations all over the world and making them accountable to the public. Public accountability is an important criterion by which presidents, prime ministers, bureaucrats and politicians are judged, so why not the CEOs of huge corporations as well? As it is, business enterprises have outgrown political institutions at the local, national and supranational level, and scrutiny by ordinary people and their representatives in NGOs is filling the vacuum. Just as with politics and the media, so it is with business: voters, readers and viewers, and now consumers, are one step ahead of most leaders.

If these are the people business depends upon, shouldn't business leaders be changing to accommodate the new demands for honesty, transparency and quality? Surely they cannot stand by and do nothing while society groans over issues that are of concern to almost everyone?

Customers want to make ethical choices and it makes sense for businesses to help them do so – though most businesses fail to provide even the most basic information. The National Labor Committee in New York, a powerful group that supports workers' rights as human rights, told me that in China it is now the law to emblazon the brand names of a company's products on its factory walls. I might add that today China produces 36 per cent of all manufactured goods for the USA, but the American corporations will not print the addresses of their factories on their labels. Why? It will 'give away competitive advantage'. Rubbish! These factories are all producing their competitors' goods alongside their own.

Personally, I do not want to buy flowers from Colombia, because I know about the diseases caused by pesticides that women are enduring in the cut flower industry. I don't want to buy from Total or Shell because of their practices in Nigeria or Burma. I just make my choices. I am opposed to maximizing profit to satisfy investors and I believe you should care for your employees and care for your suppliers. You should tell the truth to your public and your customers. Only then can you conduct your business in a profitable way.

Consumers are starting to resist – as far as they are able – the inexorable progress of the global market. By challenging the transnationals, vigilante consumers have brought about what change there has been. Some have employed brilliant 'sabotage' techniques – fabulous guerrilla tactics like buying shares in a company and then turning up and hijacking the annual general meeting.

Some just buy intelligently and demand information on the products from the CEOs or take part in consumer boycotts.

Boycotts are a paradoxical instrument to use. They threaten the jobs of ordinary people and they end the vital human connections that are needed to make change happen. But they are also very effective. Sales of Nike shoes dipped when consumers learned they were being made in Thailand by young kids in sweatshops. Nike responded to consumer criticism by promising to let NGOs monitor labour conditions at all the factories that make its products. The company has also undertaken to improve air quality and offer education and business loans to workers.

Four out of ten consumers around the world responded in some way against actions by companies they thought were unethical during 1999, according to research by PricewaterhouseCoopers, BP Amoco and Bell Canada in a survey of 22,000 people in 21 countries. The survey echoes similar research in the USA by PricewaterhouseCoopers and the Reputation Institute,

DON'T UNDERESTIMATE THE POWER OF THE VIGILANTE CONSUMER...

Anita Roddick

which showed that a quarter of the 11,000 people they asked had boycotted a company's products or urged others to do so.

Fifteen years ago, the focus of public concern was simply governments, now it's also business – from the environmental groups which are demanding change from the oil companies to the consumer assaults on major food producers and retailers of genetically modified crops. Now they are being joined by the human rights lobby, which is joining the ranks of the ethical investors, consumer boycotters and direct action specialists in spotting the so-called 'business agenda' and therefore, in my view, rightly making business the target of protest. They are also providing other people – you and me – with information they believe ought to be publicly available when we choose what and where to buy.

Faced with consumer boycotts over Ogoniland, Shell published *People, Planet and Profits*, a glossy report outlining its aim to be 'a leader in the economic, environmental and social aspects of everything we do'. This was from a company, remember, that wanted to ditch the Brent Spar drilling platform in the North Sea and invested heavily in Nigeria.

Meanwhile in Colombia, as many as 4,000 members of the U-wa tribe threatened to commit mass suicide if Occidental Petroleum subsidiary Oxy forged ahead with its plans for oil exploration in their ancestral lands. The U-wa believe drilling for oil will wound Mother Earth, creating lethal consequences for the whole of humanity. They say they weren't consulted before the project was mooted. But their protest ensured that whatever action Oxy took it would be in the harsh glare of the international spotlight, with vigilant, and vigilante, consumers ready to flex their financial muscle. Representatives of the U-wa took their struggle to the streets of Santa Monica, California, in April 2000, where they demonstrated outside the company's annual shareholders' meeting. The oil giants are learning that they can no longer hide behind consumer ignorance or indifference.

Many fear that the conversion of the likes of Nike and Shell into model global citizens is not much more than an empty public relations gesture. What is there is the rhetoric. 'We found ourselves out of synch with society,' says Tom Delfgaauw, who heads Shell's social accountability unit. 'Consumers have a choice. They can vote with their wallets.' That is certainly true.

So, the new NGOs and consumers have changed the business world. It is a world of immediate communication, of heightened scrutiny of all aspects of corporate behaviour and of growing violence as internal conflict – both the cause and effect of human rights violations – replaces interstate wars. Transnational companies have shown themselves almost wholly unprepared for these changes, although the globalization of the world economy has hugely increased their range and influence.

This movement towards ethical business is not happening because people like me say it's a good idea. It is a groundswell, a growing realization that business has to play the social role that accords with its position in our society today.

So, business leaders have a choice: they can build a huge PR wall and talk down to customers or they can listen and respond. Will they choose more smokescreens and inertia or will they listen and act? Consumers will be watching and waiting.

A Native American tribe, the Oglala, once invited me to their reservation to see if there was any economic initiative we could come up with. The Pine Ridge Reservation in the Badlands of South Dakota is the site of the Wounded Knee standoff in the 1970s, a watershed in Native American activism, but the tribe needed to survive in the world economy along with everyone else.

The first things I saw when I arrived were sage bushes, sage bushes everywhere. Sage oil is very good for the hair, so it made sense that if we could bring in some form of alternative technology, we would be able to extract the oil and then show the Oglala how to use it in hair care products. When I pitched the idea, they said they would have to 'do a sweat' to ask permission of the plant nation. So we went through a couple of sweats, six Oglala and me, packed into a tent with one man pouring water on hot rocks to generate a suffocating steam – a terrifying experience because I'm so claustrophobic.

After all that, the verdict was no. The plant nation had rejected the idea, so the Oglala did too. After my initial astonishment, I realized this was more than plain respect. It was a veneration of the natural order so ingrained as to be alien to our own *carpe diem* culture. For about a nanosecond my Western business head spun, then wonder took over. We understand respect, as in 'respect for the environment', but we've lost our sense of true reverence.

I believe we can learn something from the Oglala. As long as we can put some idealism and reverence back on the global agenda, understanding that corporations and institutions have to be a force for positive change, then there is still a light at the end of the tunnel.

There are so many aspects of life that can't be reduced to an entry in a balance book and our survival depends on remembering this. Of course this will partly mean governments and consumers remembering it, but in the end it is about business itself living up to its responsibilities at the same time as coming under scrutiny. Public scrutiny is recession proof. Companies will be judged as much on their principles as on their profitability and, as some companies have already found to their cost, it is on the basis of principle that the brightest recruits will choose where to work. Inaction is not an option. The choice is between the exercise of corporate leadership in developing appropriate company policies and being forced by public opinion to bring corporate practice into line with the values of society. The first would indicate that business recognizes that its growing influence in the world economy carries with it growing responsibilities. The second would threaten not just individual corporate reputation, but it would also jeopardize the collective licence of transnational companies to operate in the twenty-first century because their moral standards failed to match their economic reach.

BEING RESPONSIBLE

To be part of the solution means bearing responsibility for the total impact of business operations – for the way in which the employees are treated and the security arrangements are made, and for the effect of the business on the social, physical and political environment in which it operates.

This is certainly a philosophy espoused by Samsung, electronic conglomerate that is Korea's second largest company. The disastrous implosion of the Indonesian economy towards the end of the millennium resulted in a backlash of frustration and resentment against ethnic minorities and foreign companies, but Samsung managed to stay clear of the storm. At the height of the unrest, its local employees in Surabaya pulled together to protect its refriger-

ator factory and shield expatriate managers from violence. Local
workers enlisted relatives in the countryside into a food supply network that helped insulate their colleagues and even neighbouring
families from skyrocketing prices for staples like rice and palm oil.

That kind of loyalty isn't easily won. Since taking over the
firm in 1987 from his father, group chairman Lee Kun Hee has
stressed initiatives to improve the lives of local employees as part
of a broad concept of corporate citizenship. By the end of the
1990s, the firm was spending more than a third of their profits on
what it calls 'social contributions. These included undertaking Adopt-
a-River environmental schemes, building toilets, providing computer
lessons and supporting the elderly and children from low-income
families.

Business has a role in building a communi-
y that goes well beyond giving back,' says
Min Kyung Choon, who heads Samsung's
joodwill efforts. 'Charity is not enough.'

Not everyone has yet gone as far as Samsung. A bare minimum for
corporate responsibility means saying no to dealing with torturers and
despots. That is what Levi Strauss decided when it closed operations
in China. That one action, by the way, endeared the company to the
young. Indeed, the message from the streets is that consumers are
expecting moral decisions. As businesspeople, we have to rethink our
approach to these issues and then we have to act, in ways big and
small, to bring sustained and healthy growth across the globe. Our
political postures must change – we have to stop endlessly whining for
easier rules, lower costs and fewer restrictions. And our business practices must change too. We have to take longer-term views, invest in
communities and build long-lasting markets.

Less than a century ago, visionary business leaders were-
hooted out of business associations for saying that businesses had
a responsibility to support charity; they were told that the concept
of 'good corporate citizenship' was radical pap. Indeed, corporate

contributions to charity were often illegal. Depression and world wars changed us then; global poverty and environmental destruction must change us now.

There are signs of improvement. The Co-op Bank refuses to invest in businesses it views as unethical, including producers of fossil fuels and arms manufacturers. Iceland was the first major supermarket chain to ban GM ingredients and other supermarkets have partially followed their lead. But it's a slow process. So few business leaders have even begun to accept what, for me, has always been a simple truth: there is more to business than making money.

The business of business should not just be about money, it should be about responsibility. It should be about public good, not private greed.

DIFFERENT MEASURES

I believe we need to measure ourselves against a different standard. We need business that respects and supports communities and families. We need business that safeguards the environment. We need business that encourages countries to educate their children, heal their sick, value the work of women and respect human rights. Companies have to ask themselves: 'What does profit mean? Profit for whom?' Maybe we should redefine profit. We need to measure progress by human development, not gross national product.

Those measurements are beginning to count. For retailers, the pioneering efforts of Levi Strauss and The Body Shop in building human rights criteria into commercial decision-making have been underlined by the development of the Council on Economic Priorities' auditable social accountability standard, SA8000, and the UK's Ethical Trading Initiative. Both of these directives involve

My vision, my hope, is simply this: that many business leaders will come to see a primary role of business as incubators of the human spirit, rather than factories for the pro- duction of more material goods and services.

companies and NGOs in the development of principles to protect the human rights of the workforce down the supply chain.

SPIRITUAL BUSINESS

But there is more to all this than measurement and that brings us back to the word 'reverence'. There is a spiritual dimension to life that, for me, is the real bottom line. It underpins everything.

Spirituality, to me, is a very simple attitude that has nothing to do with organized religion. It means that life is sacred and awe-inspiring. In my travels around the world, I have been grounded – as millions also have – in the most fundamental of insights: that all life is an expression of a single spiritual unity. We are not, as humans, above anything, contrary to what Christianity tells us; instead we are part of everything. This interconnection has to be sacred, reverent and respectful of different ways of knowing and being.

We especially need these different ways of knowing now that the business assumptions forged during the Industrial Revolution are faltering. Predictable and controllable business environments, employees and political structures have become a thing of the past. We should be evolving into a new age of business with a worldview that maintains one simple proposition: that all of nature – humans, animals, the Earth itself – is interconnected and interdependent. We are all in this together and we are at a crossroads. We have the power to preserve or destroy the sacred interconnections of life on this planet.

It takes an enormous effort to experience spirituality within the contradictions and paradoxes of human organizations and big business. According to Krishnamurti, what we need is not just to change the system but also to change ourselves:

Systems, whether educational or political, are not changed mysteriously; they are transformed when there is a fundamental change in ourselves. The individual is of first importance, not the system; and as long as the individual does not understand the total process of himself, no system, whether of the left or of the right, can bring order or peace to the world.

I would like the big business corporations of today to learn a few lessons from the Quakers, who ran excellent businesses, yet remained utterly decent and responsible. They had a public policy, they never lied, they never stole money from the corporation and they never took out more than they put in. They cared for the community and they seemed to do really well. What is needed in business is a return to kindness and a rejection of obscenities like huge compensation packages for CEOs. I think it is a sin of the human spirit to sack thousands of people and then accept a million-dollar bonus. Maybe there should be a new word in the business lexicon: 'frugality'. It's like good housekeeping or good environmental management. Frugality is good for business in a way that profligacy is not.

The Quakers would have also supported the proposition that as business is such a powerful force in society today it ought to be harnessed to effect social change, to improve the quality of life in those societies around the world where basic needs are not being met.

What today's corporate reactionaries forget is that, long before stakeholding became a political buzzword, it was sound business practice. The great Victorian philanthropists endowed educational institutions, libraries and hospitals in their local communities, and worked hard to improve the conditions and education of their employees. They understood that a cohesive society is an essential foundation for business success and that their companies would thrive with healthier, better-educated and more productive people. It would be folly if today we didn't see the role business can and must play in the development of human beings.

I would also like to see business schools live up to the challenges of the next decade. Any future business education programme, whether set in a local or global context, must contain the language and *action* of social justice, human rights, community economics and ethics as well as the productivity of the human *soul*. The revolution is for business schools to become places where our personal values and economic interests intersect.

When I look back at the last few paragraphs, I know this is an area which evokes the most ridicule. What has it got to do with business? Everything: you have to care for and empower people to stand up and make a difference. The same goes for words like 'reverence' and 'spirituality'. Globalization is itself, in a sense, a kind of quasi-religion. The 'global market' is one of the most seductive phrases that we have yet coined. You will hear business urging us to get governments out of the way and let the markets rule. The danger is that, in the end, global markets drive everything that's valuable out of the way too. The market doesn't have a human face, a mentality or a conscience. It doesn't have a record of sympathy, shame or human endeavour. It knows neither kindness nor loyalty – and those things are essential to life.

CHANGING THE SYSTEM

I think the system can, and must, be changed – not by isolationists, but by the true internationalists. I believe that corporations operating globally can change the system to encourage trade that is fair, sustainable and devoted to good husbandry of the Earth's resources. I am persuaded that such steps will contribute mightily to political stability and true democracy.

It is high time the ideal of success should be replaced with the ideal of service.

Albert Einstei

Business should not try to be impervious to agents of change. There is much to gain from being open and accountable. Look at the way the Co-op Bank has benefited from an ethical stance. A growing number of fund managers, including Standard Life, now offer ethical unit trusts and pension funds which screen out shares in companies involved in unethical practices. Other companies, like BT, are examining new possibilities for the independent assessment of business activities. Around £4 billion is invested ethically in Britain alone. It's a learning curve for everyone.

I don't pretend for a moment that when I ran The Body Shop we were perfect, or that every one of our experiments worked out, especially when it came to building trading relationships that actually strengthened poor communities. But we were absolutely committed to thoughtfulness and sensitivity in our trade with communities around the world. The true challenge for all of us is to take on our social, and therefore moral, role in business.

ACTIVISM IS THE RENT WE PAY FOR BEING ON THIS PLANET

Over the past 23 years, while many businesses have pursued what I call 'business as usual', I have been part of a different, smaller business movement, one that has tried to put idealism back on the agenda. We want a new paradigm, a whole new framework for making business a force for positive social change. It must not only avoid hideous evil – it must actively do good.

I am not interested in business as usual. It is business as *unusual* that excites me.

ENTREPRENEURS ARE ALL A LITTLE **CRAZY** . THERE IS A FINE LINE BETWEEN AN ENTREPRENEUR AND A **CRAZY** CRAZY PERSON. PEOPLE SEE AND FEEL THINGS THAT OTHERS DON'T. AN ENTREPRENEUR'S DREAM IS OFTEN A KIND OF MADNESS, AND IT IS ALMOST AS ISOLATING. WHAT DIFFERENTIATES THE ENTREPRENEUR FROM THE **CRAZY** PERSON IS THAT THE FORMER GETS OTHER PEOPLE TO BELIEVE IN HIS VISION...

WHO WANTS TO BE AN ENTREPRENEUR?

I was brought up in a large Italian immigrant family with a work ethic that was tantamount to an unregistered form of slave labour. My parents owned the Clifton Café in Littlehampton, on the south coast of England. It was a converted Methodist chapel directly opposite the railway station. It opened at 5 a.m. to provide breakfast for the local fishermen and stayed open all day, serving non-stop plates of something-with-chips. It closed when the last customer left. I couldn't help noticing that other cafés in the town that weren't owned by immigrants tended to open at nine and shut at five. Not us!

There were four of us kids and we worked every weekend, every evening and every holiday in the café. There were no family holidays, very few family diversions at all except for the weekly trip to the cinema, and the café was an extension of our home. Courtships flourished in that café and marriages were cemented – and probably divorces too. Friendships were made and we learned a fundamental lesson about business – that it is possible to bring your heart to the workplace.

There was clearly something of the entrepreneur in my father because when I was still little he converted the café into an ice-cream soda bar, just like an American diner with a milk

Notes from my business school lecture.

bar serving exotic concoctions like knickerbocker glories in tall glasses. We all wore uniforms that resembled something out of Norman Rockwell's Soda Fountain gals and I was dazzled by it all. It taught me just how important a sense of theatre is for business and how important it is to create an atmosphere.

When I was 10 my father died and we all had to work even harder at running the café, which seated around 70–80 people and was often full. Working became an important part of our lives. It was at the Clifton Café that I learned an important lesson for all entrepreneurs. I learned that business is not financial science, it's about trading – buying and selling. It's really as simple as that. And my mother taught me how important personalities are in business. She made me realize that you can't be a nondescript person or you will have a nondescript product. **'Be special,' she would always say. 'Be anything but mediocre.'**

I also learned about trading when I was at my convent school. My father had brought back a great stack of comics and bubble gum from America. Anything American was like the Holy Grail in those days and I was able to trade the comics and bubble gum for entire collections of cigarette cards and movie albums. Although I had the whole stash at home, I pretended the stuff was arriving in small batches week by week in order not to flood the market and I would whet my customers' appetites by pretending there was an incredible Batman or Captain Marvel on its way. It was an interesting discipline.

Although I went to a Catholic school, my mother hated our local priest. She loathed him with a passion. She dressed us in trousers to go to school, but the nuns didn't like it and sent us home. She simply sent us back, still in trousers. She sabotaged Sunday mass by rubbing the hems of our clothes with garlic and squashing garlic cloves on our fingers. By doing this she could guarantee that the smell of garlic would overpower the smell of incense. When my father died, I remember sitting on the stairs of our house while my mother was furiously scrubbing the linoleum in the hallway. There was a knock on the front door and the priest was there to tell her that she was very lucky that my father was going to be given a Catholic funeral.

Mother just picked up the bucket of dirty water and threw it over him. I'll never forget that. Acts like that push you onto the edge of bravery.

It is no wonder, having a mother and an upbringing like that, that I learned to challenge everything I was told – at school, at church and in every other institution.

My mother also taught me that life wasn't any more complicated than love and work. Those were her own words.

After high school I entered teacher-training college and while I was away my mother sold the Clifton Café and opened a nightclub above a butcher's shop in the centre of the town. She called it the El Cubana and had it tarted up like a bar in Torremolinos. I arrived home to find her holding court there, sitting at the bar smoking – she had never smoked in her life – and wearing a silver Lurex dress. The set design of the place was perfect – tacky, but it worked.

I've also tried to create atmosphere, whatever I've done. When I started teaching, I incorporated drama and music into my lessons whenever I could. If I was giving a lesson on medieval history I would play Gregorian chants and if it was the First World War I'd read the war poets.

Although I enjoyed teaching, I had itchy feet and I took off on the hippy trail for a while before talking my way into a job with the United Nations in Geneva. I knew that if I wrote in for a job no one would take a blind bit of notice, so I just showed up on the doorstep and asked to see someone in the personnel department. I had virtually no qualifications but a great deal of confidence in my ability to sell myself and I literally talked them into giving me a job. Energy and enthusiasm either daunt people or seduce them. Luckily for me, the UN was seduced.

GOING INTO PARTNERSHIP

After two years of travelling around the world, I returned home and my mother informed me that one of her customers at El

Cubana was longing to meet me. It was Gordon Roddick. Four days later, I moved in with him. Gordon had trained as a farmer but was trying to earn a living writing children's stories. I got a job as a teacher in a junior school, but I was soon pregnant with Justine, our first daughter. After she was born, Gordon worked as a labourer to earn more money. We talked a lot about starting our own business but never seemed to be able to make a decision about anything.

When Justine was 15 months old, I found I was pregnant again and we decided to take a trip to the United States to visit friends. We were supposed to be looking for 'business opportunities', but actually we were just having a good time. One day we went out for a drive, found ourselves in Reno and decided on the spur of the moment to get married. The ceremony was conducted with Justine howling in a harness on my back.

After Samantha was born in July 1971, we started looking for a business in earnest, something that we could run together and that would still leave us time to look after the children. There was a run-down Victorian house for sale in the centre of Littlehampton. It was a former residential hotel which had fallen on hard times, but we thought it had the potential to be turned into a fantastic little bed-and-breakfast place. We made a silly offer and to our amazement it was accepted. We refurbished it very cheaply and within a few days of opening up for business we were doing a roaring trade.

Everything was going fine until the summer season ended. Suddenly there were no more customers and we found ourselves saddled with a huge empty house and horrendous maintenance costs. We realized we were going to go bankrupt unless we did something and did it quickly. What we did was to turn half the house back into a residential hotel. That saved us and taught us a lesson valuable to all budding entrepreneurs: when you make a mistake, you have to face up to the fact and take immediate steps to change course.

Once the hotel was running smoothly, full of confidence, we decided we would open a restaurant as well. We borrowed £10,000 from the bank to buy the lease on premises not far from the hotel and set about creating a very stylish restaurant called Paddington's, with bentwood chairs and potted palms. Our idea was to offer healthfood dishes with an Italian flavour – lasagne and quiches and homemade soups.

It was a disaster. No one seemed to share our enthusiasm for the menu. Day after day we sat in an empty restaurant and within three weeks we were on the verge of going broke again. What saved us was our ability to recognize that we had done everything wrong. It was the wrong restaurant in the wrong town. We were trying to impose our will on the customers and sell gourmet food in an egg-and-chip environment. It was Gordon's suggestion to change tack, get in a deep fat fryer and a chargriller and turn the place into an American-style hamburger joint with loud rock music. The effect was miraculous. Almost overnight Paddington's became the most popular place in town.

THE START OF THE BODY SHOP

After three years of running both the hotel and the restaurant, we were exhausted and we decided to get out. Gordon had a long-term ambition to undertake a horseback expedition from Buenos Aires to New York and while he made plans for that we decided that I would make ends meet with a small shop of some kind. I didn't want to go back to the gruelling business of running a restaurant and thought that a nine-to-five shop would provide an income while allowing me time to spend with the children.

I already had an idea of what kind of shop I would like. Skincare would be so easy. You could find out everything by reading, researching and conversation.

My mantra was: 'Make the past into a prologue for the future.'

I was trained as a history teacher so I knew how to conduct research, where to look, how to dig deeper and deeper. I read every pharmacopoeia book from the turn of the century onwards – books on grandmothers' secrets, books on kitchen cosmetics. I yelped whenever I came across useful anecdotes. Julie Christie liquidizes cooked lettuce and avocados for a face cream. I did it. Marlene Dietrich collected carbon from candle flames and used it as eye shadow. I did it. Whatever I read I used. It wasn't about business, it was simply about creating a livelihood – *my own* livelihood. It was about being mistress of my own time and space, redefining my own success as a sense of freedom.

In the cosmetics industry at that time you had no choice of sizes, so you were browbeaten into buying what you didn't really need. It seemed ridiculous to me that you could go into a sweet shop and ask for an ounce of jelly babies, but if you wanted a body lotion you had to buy a much larger amount. Also, everything was expensive and very élitist and there wasn't anyone selling natural ingredients. This irritated me. If something irritates you, it is a pretty good indication that there are other people who feel the same.

Irritation is a great source of energy and creativity.

I sat down with Gordon and told him the kind of shop I was thinking of opening was one that sold natural cosmetics in different sizes and in cheap, refillable containers. It made perfect sense to him and anyway we knew from our experience with the hotel and restaurant that if my idea didn't work we could keep tweaking it until it did. If we couldn't sell creams and shampoos, we'd chuck them out and sell something else.

The first branch of The Body Shop opened in Brighton in March 1976. Everything was determined by money, or rather a lack of it. I hired a designer to come up with the logo for £25 and I got friends to help with filling the bottles and handwriting all the labels. I painted the whole place dark green, not because I wanted to make an environmental statement – the word 'green' was not a metaphor for the movement then – but because it was the only colour that would cover up all the damp patches on the walls. The cheapest containers I could find were the plastic bottles used by hospitals to collect urine samples, but I couldn't afford to buy enough so I thought I would get round the problem by offering to refill empty containers or fill customers' own bottles. In this way we started recycling and reusing materials long before it became ecologically fashionable. **Every element of our success was really down to the fact that I had no money.** I ran my shop just like my mother ran her house in the Second World War — refilling, reusing and recycling everything — and what we did in that first year was a thumbprint for the differences that would set the company apart.

BEING DIFFERENT

From the very beginning we wanted to be able to tell stories. We wanted to be honest about the products we sold and the benefits they promised. We wanted to go in the opposite direction to the rest of the cosmetics industry. I would give that advice to anyone: set your sights and skills on an idea, do the research, see what the competition is doing and then see how you can be different. Focus on what the competition doesn't have and promote that.

My thinking was forged in the 1960s and in those days I would rather have slit my wrists than work in a corporation or even consider a business career. So we had no organizational chart, no sophisticated procedures or marketing, no one-year, five-year or ten-year plan. In fact we didn't have any of those things for the next 17 years. What we did have was management by our common values, but we were also in some ways falling apart at the seams. **When a company grows, it is like watching your child grow. Everything it does is exciting. When it stumbles, says its first word, it's all exciting.** Then it grows into this teenager and then into a mature adult and maturity presupposes an unwillingness to change. How do you get a mature organization to act in a delinquent way, to keep that sense of entrepreneurship, that sense of derring-do? That became the challenge later and it is terribly hard, because managing sameness deadens the soul. I didn't want to be the same as everybody else or even be the same as last year. And as for managing, I would rather be faced with trying to achieve harmony and goodwill among creative people who are at one another's throats than with trying to squeeze an ounce of innovation or creativity or risk out of a company full of clones. But then, I don't think that's unusual in an entrepreneur.

BEING AN ENTREPRENEUR

As the founder of The Body Shop, I often get asked to talk about entrepreneurship – even by hallowed institutions like Harvard and Stanford. It makes me smile that the Ivy League is so keen to 'learn' how to be an entrepreneur, because I'm not at all convinced it is a subject you can teach. How do you teach obsession, because more often than not it's obsession that drives an entrepreneur's vision? How do you learn to be an outsider, if you are not one already? Why would you march to a different drumbeat if you are instinctively part of the crowd?

I never set out to be an entrepreneur, I'd never heard of the word and I was not interested in its definition. But since those early days I have had plenty of experience of the ups and downs of entrepreneurship and I've met many other entrepreneurs I have liked and admired, so I feel I can discuss the subject with a little authority. After a quarter of a century trying to reinvent business, I've come to the conclusion that the qualities you need to be a natural entrepreneur include a combination – at least – of the following:

1 The *vision* of something new and a belief in it that's so strong that it becomes a reality. Vision-making is also obsessive, a type of psychopathology. It is inherently crazy. If you see something new, your vision usually isn't shared by others.

2 A touch of *craziness*. There is a fine line between an entrepreneur and a crazy person. Crazy people see and feel things that others don't. An entrepreneur's dream is often a kind of madness and it is almost as isolating.

3 The ability to stand out from the crowd because entrepreneurs act *instinctively* on what they see, think and feel. And remember there is always truth in reactions.

4 The ability to have *ideas* constantly bubbling and pushing up inside until they are forced out, like genies from the bottle, by the pressure of creative tension. But all these ideas are nothing, of course, unless someone can expedite them, which is where you thank God, or the gods, or both, for the people who have that skill.

5 Pathological *optimism*. Everything is possible for an entrepreneur. This extraordinary level of optimism bears no relationship to any degree of planning.

6 A covert understanding that you don't have to know *how* to do something. Skill or money isn't the answer for the entrepreneur, it is knowledge: from books, observing or asking.

7 *Streetwise* skills. Most of the entrepreneurs I've met have had an innate desire for social change. They understand that business isn't just financial science, where profit is the sole arbitrator, it is just as much about taking part in political and social activism, using products as conduits for social change. That gives entrepreneurs enormous freedom to experiment with what they want, but it also makes them dysfunctional in hierarchies and inert structures.

8 *Creativity*. Of course it's easy to talk about creativity, but in essence it remains a mystery to me. I have never heard or read anything that explains how people behave creatively, despite the fact that we constantly glory in human creativity. Einstein said: 'Imagination is more important than knowledge.' Dali claimed: 'You have to systematically create confusion, it sets creativity free.' Maybe creativity is magic, maybe it is bestowed by the gods, maybe it is just polished opportunism. I just don't know and I'll probably go to my grave not knowing.

9 The ability to *mix* all these together effectively. For me, becoming an entrepreneur was a consequence of simply trying to blend the skills I possessed into creating a livelihood. I learned by experience. So I don't believe you have to go to college and study at the feet of some nutty professor of entrepreneurship. I think you have to ask questions of everyone, and never stop asking questions, and knock on doors to seek as many different opinions as exist. Then you have to make up your own mind and plough your own furrow. I have never read a book on economic theory or business theory and I don't intend to. It's not theories that interest and excite me – it's the doing that keeps me going.

10 And finally, every entrepreneur is a great *storyteller*. It is storytelling that defines your differences.

Entrepreneurship achieved cult status in the 1980s and then became absolutely *de rigeur* in the 1990s, but through the past two decades we have rarely heard anything about the danger of over-glamorizing this category of human beings, or, for that matter, anything about what makes us so crazy.

One of the great challenges for entrepreneurs is to sit down, reflect and wait to collect information. We all suffer from hurry sickness. We have an abundance of energy and commitment. We can create something out of nothing, but we're not very good at organization and follow through. We are the kind of people who 'march to a different kind of drumbeat', who do not see ourselves as part of the mainstream. We are essentially outsiders and that is the best definition of an entrepreneur I have ever come across.

THE OUTSIDERS

Even assuming that it is possible to teach entrepreneurship, I think it would be especially hard to teach the rich. The rich are very rarely outsiders. You might be able to teach them the 'science' of entrepreneurship, but you can't teach juicy, backs-to-the-wall, seat-of-the-pants, bootstrapping stuff to people who don't know what

Leading by inspiring, support-ing, and the employment of other so-called feminine qual-ities will be the new style of managing. Willis Harman frugality is, who don't have a passion for wanting to establish freedom, freedom from structure and process. And creativity, the very essence of entrepreneurship, is often stimulated by being hard up. If you are broke, you're hungry. If you are wealthy or middle class and everything is available without a struggle, you don't have the hunger that drives entrepreneurs.

You've got to be hungry – for ideas, to make things happen and to see your vision made into reality.

Surprisingly many entrepreneurs have known deprivation. There is often a loss in their lives, perhaps the loss of a childhood by being forced to work at an early age, or being sent away to boarding school, or losing a parent, or finding themselves thrust onto their own resources by circumstances out of their control. In my own case the death of my father was totally traumatic for me.

Entrepreneurs are outsiders by nature – outsiders with a work ethic. That's why immigrants make such good entrepreneurs: they are outsiders but they are not frightened of work. The Bangladeshi community in Britain, the Jews and the Italians are not part of the throng; they look at things in a different way. They are not scared of sacred cows like bank managers or hierarchical structures or problems that would daunt others.

Nothing frightens true entrepreneurs because nothing can be allowed to interfere with their vision. They have the same passion as artists and writers. Just as an artist creates a painting from scratch, so entrepreneurs can realize their own dreams in precisely the same way – by turning an idea into reality, earning a livelihood from it and, hopefully, making a profit.

Any idea you have is like an extension of your personality and any company that you build yourself has your thumbprint on it. It's an extension of yourself.

The macho image of leadership, associated with men like Vince Lombardi, Ross Perot and Lee Iacocca, makes us forget that the real strength of a leader is the ability to elicit the strength of the group.

Richard Farson

WOMEN ENTREPRENEURS

Is it harder for a woman to be an entrepreneur? Certainly the enterprise culture is full of contradictions for women. On the one hand you are encouraged to get out there and conduct your own destiny, but on the other hand there is the very strong moral authority of home and hearth – the idea that we should be where our kids are. It is still far easier for any woman to go to a bank and secure a loan for a new kitchen or fitted wardrobe than one for starting a business. There's still the prevailing notion that women don't have the necessary business skills.

what's a nice girl like you doing in a place like this?

i own it.

SEXISM'S TOAST

Another cliché is that women bosses run a caring, sharing ship. According to the results of a study by Manchester Business School, the truth is otherwise. The survey says women at the top are tough cookies, autocrats who rule by fear. I wonder if that isn't a misreading of the entrepreneurial spirit, regardless of gender. People who are good at starting something are not so great at running it. For that reason, I have always felt that just about the best thing I can ever do in business is find and employ people better than myself.

In fact there isn't much information available about women entrepreneurs, so it is difficult to establish exactly what the differences are. I remember a government official explaining that data isn't collected on the differences between male and female-run business because that would be 'sexist'.

Entrepreneurs are far more likely than corporate managers to intertwine their business and personal lives.

Marketing News

One huge difference, of course, is that women's work is often unrecognized. Where are the studies that show how women have positively influenced the workplace, made it more productive with their relationships and their intimacies? For that matter, where are the studies on the invisibility of women's labour?

The sad thing in this country is that women in business don't talk to each other enough, whereas in America women share skills and network brilliantly. Americans are phenomenal entrepreneurs, they are great at making connections and they have a real respect for women entrepreneurs. I also think they are naturally better at taking risks and coming up with new ideas: they are more opportunistic and optimistic. Small-business entrepreneurs are the backbone of America's economy – they employ more people than the Fortune 500 companies. Here, in the UK, we're still narrowing the gap between what men value and what women are worth.

But despite all this, I believe that women have an entirely different approach to entrepreneurship than men. Women tend to balance what they are interested in and what they are enthusiastic about and are perfectly happy to trade that into a very low-tech business, often operating from the dining room or kitchen or a garage. Men, in my experience, want to shape their ideas into much more of an operation. They tend to want all the trappings of business, like the car, the mobile phone, the office and the secretary. I don't know whether that makes them better businesspeople – I rather doubt it. I certainly think women are much better jugglers; they deal with complexity better – looking after kids, running a home and running a business at the same time. I don't think there are many men who could do that.

If a woman can decide who gets the last toffee, a four year old or a six year old, she can negotiate any contract in the world.

What separates the entre-preneur from others is that entrepreneurs act on what they see. William B. Gartner

Unfortunately, unlike men, women are consistently judged by their appearance. I am often aware when I am giving a speech that the audience first weighs up my appearance before listening to what I have to say. That would never happen to a man. What is even more irritating is that the way women look is still frequently given greater weight than what they are saying. It is all part of this ludicrous culture of seduction. So we're damned if we look good and damned if we don't.

When I went to a bank to try and get a loan to open the first branch of The Body Shop in 1976, I turned up with both my children, wearing jeans and a Bob Dylan T-shirt. It just didn't occur to me that I should be anything other than my normal self. **I explained that I'd got this great idea for a shop and all I needed was £4,000. I was turned down flat.** A week later I went back with Gordon and saw the same bank manager. We were both wearing suits and this time I had an impressive-looking business plan bound in a plastic folder (actually it was a lot of gobbledegook). We got the money with no trouble.

I don't think many women would go about it the way I did, but I just didn't think about it. **I expected to be judged on my ideas and my values, rather than what I looked like, but I was wrong. Today, my advice to any woman in a similar situation would be to leave the kids at home and dress anonymously. Don't, whatever you do, look the least bit untraditional.**

ADVICE TO ENTREPRENEURS

So here's my advice to entrepreneurs. The first thing is:

Be opportunistic. Successful entrepreneurs don't
work within systems, they hate hierarchies and structures and try to destroy them. They have an inherent creativity and wildness that is very difficult to capture. But they have antennae in their heads. I can walk down the street anywhere in the world and I have always got my antennae out, evaluating how what I am seeing can relate back to The Body Shop, whether it be packaging, a word, a poem, even something in a completely different business. I find myself saying: 'How could this relate to us, how could this work for us?'

When I was in a market in India, I saw all these stainless steel cans for carrying water. I looked at that volume of shapes all in the same material and thought what wonderful packaging they would make. San Francisco is famous for its sour dough bread. When I was last there I began wondering what it would be like if you made a raw sour dough mixture and washed your body with it. Then I began experimenting with balsamic vinegar. I knew it was good on the hair but I wanted to see whether it could be similarly good on the skin. I am always thinking of things that might go on the hair and body, looking at different textures, different ideas. So, tap the energy of the opportunist.

The second piece of advice is this:

Be passionate about ideas. Entrepreneurs
want to create a livelihood from an idea that has obsessed them. Not necessarily a business, but a livelihood. Money will grease the wheels, but becoming a millionaire is not the aim of an entrepreneur. In fact, most entrepreneurs I know don't give a damn about the accumulation of money. They are totally indifferent to it – they don't know what they earn and they don't care. They don't care about the nameplate culture. What gets their juices going is seeing how far an idea can go.

Being an entrepreneur is not within the realm of rational discussion. It is a burning, sometimes pathological need. Moses Znaimer

FAVOURITE ENTREPRENEURS

You can see this combination of qualities in some of the most ingenious and creative entrepreneurs around the globe. I went to Kathmandu during the trade wars between India and Nepal. There was no oil coming into the country and no taxis about, so some fast-thinking locals had put planks of wood on wheels and turned them into makeshift 'taxis'. It hit me immediately that that is exactly what entrepreneurial thinking is about. Entrepreneurs see opportunities that others don't. If there is a problem, while other people are just moaning, entrepreneurs will say: 'OK, what can I do about it?'

Take people like **Yvon Chouinard**, the founder of Patagonia. He was just a wild eco-mountaineer who was angry that the clothes he wanted to wear for mountaineering weren't available, so he made them himself. His company is now one of the leaders in the socially responsible business movement. I really like entrepreneurs like that. They are making a product, providing a service and doing good.

My all-time favourites were **Ben & Jerry**, the two guys who started up probably the world's most famous ice-cream brand, Ben & Jerry's. Neither of them knew much about ice cream apart from the fact they liked to eat it. They just wanted to play and have fun, so they took a $20 course in how to make ice cream and built an international business. They would have been murdered at business school – anybody like that would have been murdered. But they had a good idea, had good hearts and recognized that profits and principles can go hand in hand.

MANAGING SUCCESS

How do you keep a fearless entrepreneurial spirit when your business gets institutionalized and its success gets institutionalized? As I said earlier, this is the hardest part. Managing success seems to kill the entrepreneurial spirit.

What every entrepreneurial company needs is crazy people reshaping strategies, coming up with ideas, constantly experimenting. For an entrepreneur to maintain a sense of entrepreneurship, there should be half a dozen pockets of small-scale experimentation going on all the time. I used to have two or three shops in England where to some extent I experimented with products and visuals and watched what happened. That was joyful. That's experimentation at its best, when it is covert and when you can beaver away, polishing the experiment, shaping it until you have developed the idea. Then, and only then, you grin and move on.

THAT VITAL ENERGY

I measure people against myself, judging whether they've got the same enthusiasm and energy. The yardstick is the creative time they put in. It may be a wrong yardstick, but I loathe the nine to five attitude. I don't care if people are slightly chaotic. If they can motivate, that is the important thing.

People say I have an over-abundance of energy and the British tend to find that very disconcerting. They don't quite know how to deal with it. Somebody once called me 'the Blur'. A very short-legged blur! But where does it come from? I think it comes from four places:

Panic

I believe we only have one chance. I don't believe that when I die, there is a place or a journey to somewhere else. And there is no dress rehearsal in life.

You hold on to life's torch, and you keep it as bright as you can, and then you pass it on. And I think that is where the energy comes from.

Being an Outsider

It also comes from being part of a family that was always on the outside. Immigrants were always marginalized. I remember hearing a wonderful tape of Isabel Allende talking to Alice Walker about people who are marginalized for whatever reason, whether family, birthright or a strange viewpoint on life. Being the only Italian immigrants in a little blue-collar town like Littlehampton gave me the ability to observe what I didn't want to be.

A Sense of Outrage

I think discontentment drives you to want to do something about it. And my outrage came very early on.

Tomatoes

Finally, it's because I'm Italian and Italians eat a lot of tomatoes. Maybe there's an enzyme in there that makes you very perky, because the Italians I know never seem to go through real 'downs'. My guess is it's to do with tomatoes.

THIS IS NOT THE WAY I WANT TO DIE, HEAVING INTO A LAVATORY BOWL  IN KATHMANDU. THERE'S A MANTRA IN MY HEAD:

'THE WORLD IS UGLY, THE PEOPLE ARE SAD.'

WALLACE STEVENS, THE AMERICAN POET, SAID THAT.

NAH!

THIS BATHROOM IS UGLY AND I'M SAD. SAD THAT AFTER 20 YEARS' TRAVELLING, I'M BLOODY DAFT ENOUGH TO EAT PRAWNS IN KATHMANDU.

WE WERE SEARCHING FOR EMPLOYEES, BUT PEOPLE TURNED UP INSTEAD

My first real lesson about what community means, apart from the café in Littlehampton, was at a kibbutz. I won a scholarship to study the children of a kibbutz for my education thesis. It was extremely hard work, but an absolutely seminal experience. I had arrived from an academic institution where it was simply assumed that educated people were superior because education improves people and makes them good. So I carried around a set of preconceptions that non-academic work was a kind of alternative to play, that work was both the burden and the privilege of our species and that it was a joyless activity. But in the kibbutz in Israel, I learned that work could be a fine thing.

I worked from around 4 a.m. until 11 a.m., by which time we were exhausted and it was too hot to carry on. My experiences in the kibbutz taught me about the value of work – about people who not only work for the common good, but who also know how to live with and connect with the land. I learned that love, labour, community, service and the land were all interwoven – and that to be physically exhausted by honest labour (so I believed then) was rather a noble experience.

I also learned that a community provides more intimacy than our normal divided lives and that the health of a community

My notes for an article in the *Independent on Sunday* about a trip with a film crew in 1997 to look at a community trade project in Nepal.

depends absolutely on trust. A community knows itself in a way that is impossible for a nation or a state. It holds marriages and families together. Families will help each other and privacy will be expected between them. There is no alternative in a community but to commit to a common purpose in a culture of participation. It has always been a sadness for me that I never lived my life in a commune.

You can learn a great deal about community by looking around you – in fact, it's something you can only learn about from other people.

We have learned a great deal over the years from other companies with the similar philosophies of community to The Body Shop. Yves Chouinard, the founder of Patagonia, is setting an incredible example in terms of political activism. Patagonia pays the bail for any employees who are arrested during demonstrations related to the environment. The company is committed to abolishing pesticides and is totally focused on its products. Patagonia is making clothes from recycled bottles, for example. Ten years ago, everyone was doing that, but when it stopped being sexy Patagonia, uniquely, stayed with it.

Patagonia's 'statement of purpose' lays down that the company 'exists as a business to inspire and implement solutions to the environmental crisis'. It funds groups that protect and restore wilderness and biodiversity. It campaigns aggressively to force governments and corporations to clean up their own mess. It has also given more than $14 million to 900 grassroots environmental groups and encourages employees to work for non-profit organizations by giving them leave of absence on full pay for two months to work for the environment.

Patagonia has been a consistent source of inspiration for me. When I first visited their headquarters more than 20 years ago I was hugely impressed by their childcare system. They didn't just have a crèche, they had a 'child development centre' – and the words were enough to see their ambition. Often if you give a project the right name, you end up working to that vision. Within a year, we

Community comes from the word communion, to share a common task together. And it's in the sharing of that task that people do bigger things than they knew they were capable of. Then there's really something to celebrate. Matthew Fox, theologian had set up a similar unit at The Body Shop, the first one in the UK attached to a workplace, as far as I am aware.

I have also learned a lot from my children. My younger daughter, Sam, lived in Vancouver for many years, in a place called The Drive. The Drive is a community of refugees, writers, musicians and artists, most of whom are out of work, and it has a culture of 'participation'. People who live on The Drive are not onlookers, they are activists. They see no alternative but to commit themselves to a common purpose. They told me they have a 'war room' in an old furniture depository in which they have maps of every street in Vancouver where dumpsters are positioned, maps showing which dumpster holds which goods, whether they are perishable, electrical, ironmongery, old refrigerators or whatever, and they have a system of taking out those goods, cleaning or refurbishing them and then distributing them throughout the community. It is different from the way other people live, but it gives them something they can't find anywhere else – the feeling of belonging, of erasing loneliness and of commitment.

When I looked at my daughter and the community in which she lived, with its alternative education and alternative economics, I felt I was looking at a picture of what was happening within the 'community movement' around the globe. When I spent time with Sam, I saw what you need to sustain a successful community: personal commitment. Sam and the people she lived with embodied this.

Over the past 30 years, I've also come to realize that this sense of community is absolutely vital to business success. But it isn't simple. There are communities to be nurtured on at least at three levels: inside your company; the wider global community to which any company owes certain responsibilities; and the specific communities with which you trade.

All this talk of community sometimes seems completely alien to modern business, with all its financial projections and money measurements. Yet the concept lurks behind everything. The difficulty for entrepreneurs who want to be rooted in community is to measure that success in a way that is recognized by the financial community. That issue is at the heart of this chapter. In some ways, the domination of this kind of measurement by the

One of our greatest frustrations at The Body Shop is that we're still judged by the media and the City by our profits, by the amount of product we sell, whereas we want, and have always wanted, to be judged by our actions in the larger world, by the positive difference we make. The shaping force for us from the start has not been our products, but our principles.

financial markets is a sort of tyranny, a financial fascism. Because that narrow perspective can be quite literally soul-destroying. I've often wondered what has protected my soul over the years in a business environment that usually alienates humanity in every way. Here are some of the answers:

☞ We didn't know how to run a conventional business. We had never read a book on economic theory and had never even heard of Milton Friedman.

☞ We valued and respected labour as fuzzy and cuddly, nerdy as that sounds. We understood that life was no more complicated than love and work.

☞ We had no money. Every original idea was based on reusing everything, refilling and recycling everything we could.

☞ We were naïve. We didn't know you could tell lies. That grace has stayed with us to this day.

☞ We loved change. We believed everything was subject to change.

☞ We had a secret ingredient called euphoria. We shared an extraordinary level of optimism, and we still do.

☞ Finally – and this was the main ingredient – we couldn't take a moisturizer seriously!

I also discovered, very early on, that I love retailing. I don't like systems, financial broadsheets, analyses, three-, five- or any-year plans, but I love buying, selling and making connections. I love the productivity and the creativity. I love the values. I can look at retailing and merchandizing and it gives me a better understanding of society. Think of those open-air markets in Mexico or Ghana. Think of people standing in line for an allocation of bread in the Soviet Union. Think of those little corner shops in the 1930s, but also the mega-malls in the US with dozens of stores, potted plants and merry-go-rounds. And think of loneliness, the new religion they call consumerism and the hidden persuaders of advertising...

When I started The Body Shop, work for me was about livelihood. It was an extension of my home, my kitchen. Courtships flourished in my shop in Brighton, friendships connected, just as they had in the Clifton Café. The shop taught me that business is about buying and selling in that magical place where buyers and sellers come together. In our case, it is about the shops and about making a product so good that people will pay you a profit for producing it. **That's business. It's all about community – and I don't believe this definition really encompasses people who sit in front of computer screens, moving millions of dollars from Japan to New York.** I promised myself that if The Body Shop ever became like one of those dime-a-dozen cosmetic companies using the same language and spreading the same lies that are perpetuated by the many beauty businesses, then I would leave it like a bat out of hell.

So we are working overtime to find a fresh language, a spiritual dimension to the workplace, new boundaries.

The challenge for business leaders in the twenty-first century is to assume the mantle of spiritual elder for their cultures, so that life doesn't become trivial and grey for all the people who spend most of their life at work. **Jim Channon**

The old views of business as a jungle where only the vicious survive will, I hope, soon be giving way to a new view of business as a community where only the responsible will lead. If your values are heralded and if your heart is in the right place, if your feelings are recognized and your spirit at play, I believe there will be footprints out there for all of us.

COMMUNITY WORK(S)

COMMUNITY AS THE COMPANY

I learned a fundamental truth soon after the first branch of The Body Shop opened. It is this: when you advertise for employees, surprise, surprise: *people* come instead. They are people with aspirations, people who do not want to leave their values at the workplace door, people who want to be able to work without abrogating their beliefs, within a community of the workplace.

This revelation transformed our consciousness at The Body Shop. I think we all realized that work must be imbued with opportunities for personal growth and discovery. It was and still is our job as managers in these new kinds of businesses to provide an environment where the hearts and minds of employees can grow. Employee motivation wasn't nearly so important in the past, but these days people expect more than simply earning a crust for getting the output right. They dream bigger dreams.

The workplace for me has always been a community, a place where people work for the common good. It should also be a special place, creative, fun and an incubator for the human spirit, like the Child Development Centre attached to our headquarters in

Littlehampton. I believe that the workplace has to be where parents are served, child development needs are supported and families are welcomed and valued, explored and protected. It is also a brilliant excuse to develop young eco-warriors. We have about 60 kids under the age of five in our daycare centre and another 30 up to the age of 12 go to our day camps during the school holidays.

COMMUNITY AS THE GLOBE

As I have travelled around the world, I have come to understand that all people have similar aspirations. We all need to be loved, and we all need the same basic human rights like food, shelter, meaningful work, education and a spiritual community.

Many businesses, including The Body Shop, are now part of the social responsibility movement, a dynamic and growing movement that is trying to put idealism back on the business agenda. It's not new. In Britain it goes back to Robert Owen, the early co-operative movement and to the Quakers. In the US, the Amish, the Shakers and scores of other, frequently religious, communities have used these guiding principles to run their businesses for decades. These principles are now re-emerging in our corporate consciousness.

The social responsibility movement received a great impetus from the counter culture of the 1960s, the activist movements and the progressive business practices in Scandinavian countries. We are now successfully challenging the shibboleths of the chambers of commerce, turning instead to another agency – the Businesses for Social Responsibility group. We are networking and exchanging best practices and ideas. We are creating new markets of informed and morally motivated consumers. We are succeeding and thriving as businesses and as moral forces because of the gritty determination of the people involved.

If you want a practical definition of a socially responsible company at the beginning of the new millennium, I'll gladly offer up The Body Shop: survival by the skin of our teeth, hanging on to values while being pulled apart by internal and external pressures, yet still moving forward into something bigger and braver.

We believed you couldn't add value unless you have values to add, that's why our concepts of quality and value had extended beyond the products into the ways the profits from those products were handled.

There are obvious reasons for the success of The Body Shop – the products themselves, of course – but I also believe that if companies are in business solely to make money, no consumer can fully trust what they do or say. **With us, people respond to a redefinition of business, where the human spirit comes into play.** The social responsibility movement is complex and often paradoxical, not least because the definitions are so ill-defined. Is it enough just to practise good environmental housekeeping, clean up your mess, be kind to your employees or care for the community you're in? Is it enough to be transparent about finances? I believe it is those things, but it is also more than that. The trouble is that in communicating these paradoxical and unfamiliar ideas to the wider public, they tend to be reduced to a basic mush, with no shades of grey, leaving just black and white, good or evil.

Whatever the label – 'socially responsible', 'socially responsive' or 'socially reflective' – we are showing that business must be a force for positive social change. We have to legitimize the movement, because if in the next decade we can shift seamlessly from business

However expressed, values carry the message of shared purposes, standards and conceptions of what is worth living and striving for; and they have immense motivating power. John Gardner as a vehicle for private greed to a vehicle for public good, then there will be genuine cause for celebration.

I'm not asking companies to be absolute paragons of virtue – though I would like masculine-controlled, obsessive notions of the process of management to give way to more inclusive and feminine forms of collaborative and informal networks, and for business to take responsibility for the whole. The business world is dynamic, but it's not a black and white world, with perfect ethical companies at one end and old-fashioned shareholder capitalists at the other. We all live and work somewhere in between. It's just that I hear much about the need for increased growth, but little about stronger communities. I hear much about the march of progress, but little about the people and cultures being trampled underfoot. What the world desperately needs now is to have more value added to it, and that's what business should be about. **It is difficult to be a values-led business. I have found that people who harbour dreams of a more compassionate world often feel alone and unsupported, except when something occurs to bring their feelings out in the open. Then they find that others have also harboured these same, apparently subversive thoughts and feelings.** The passion may be isolating, but it is also very easy and spontaneous. Actually putting it into practice is far from simple – guaranteeing the integrity of our chain of 2,000-odd suppliers, making sure that any ingredient that we buy hasn't been tested on animals, that any material we buy is really renewable, that any wood has been certified, that nothing we buy is used to prop up a regime that offends human rights. It is a complicated business. And I have learned, over the years, that our people consider the social and environmental values as part of the DNA of the The Body Shop. You mess with it at your peril.

Yet the bottom line is that The Body Shop has its integrity intact. We are honest about our methods and our mistakes. We are not perfect – it isn't possible to be perfect – but we are trying to go in the right direction and in those circumstances, it's best not to mystify what we are trying to do. It is all too easy in business to be distracted by profits, the technology, the cost-effectiveness, the delivery systems. What is important is to never lose touch with what lies at its heart and soul – to remember why you are doing it in the first place.

COMMUNITY AS TRADING PARTNERS

We used to buy millions of soap bars from a European supplier. The quality was good and they were cheap, but we learned that they were cheap because our supplier was using cheap immigrant labour. This directly contradicted so much of what we stood for and offended so many of our principles that we decided we had to make a change. It also presented us with an opportunity to help a community in distress. We built our own soap factory.

We could have set it up in a safe suburban industrial park, but I would rather employ the unemployable than the already employed, so we located it in Glasgow, in Easterhouse, an area of extreme deprivation and, what is more, in our own backyard. It was a moral business decision that worked. **And why do these moral business decisions work? Because consumers understand that their purchases are moral choices as well.** If anyone was to ask me what I was most proud of in the years I was running The Body Shop, one example would be the soap factory, called Soapworks. Easterhouse was considered one of the worst housing estates in Western Europe. Now Soapworks employs more than 120 full-time permanent employees, who make some 22 million bars of soap. Not only is it independently managed, but profits are allocated back to the community. I'm a firm believer in small-scale economic initiatives and Soapworks is a micro-enterprise that has succeeded in regenerating its surroundings. Its success makes me even more committed to the campaign for community-based businesses.

Some of the biggest successes of The Body Shop have been our dealings with poor communities. We build trading relationships with poor communities with the express purpose of raising, not lowering, the quality of their lives. We did it in Glasgow, but we also do it in India, in Africa and in the Americas. **By them-selves, these initia-tives will not transform the global economy, but they have transformed the company's thinking about its responsi-bilities as a business.** The unmet challenge of Western institutions is to recognize the importance of taking responsibility and building new models of progress.

For me it's also about reaching out to other organizations and corporations. Where you connect with others there is deep thoughtfulness, and sometimes intense debate. I'm not just talking about the trading associations here, I'm including alternative trade federations, human rights groups, the co-operative movement, the community economic development ideas, independent think tanks, radical academic philosophers, economists and political scientists. This informal democratic network, this process of connection, is vital to our ability to shape a deeper thoughtfulness. It would be very dangerous to dismiss this part of the movement because of its occasional anarchy, because it is precisely this which is fashioning a new order of consciousness and framing the bigger questions: how to create human fulfilment or mass work, how to tackle social alienation or the stranglehold of the multinationals on the global economy.

THE RESPONSIBILITY REVOLUTION

You can't stop business from going global, but you can make it listen to the responsibilities that go with jumping onto the globalization bandwagon. I don't think this can be achieved by government regulation. It comes from the businesses involved finally seeing that acting responsibly and responsively – there is a difference – is actually *good* for business. Business cannot avoid moral choices, after all: its future depends on it.

It also can't avoid community, because its future depends on that too – even if it's just to educate the next generation and bind society together enough to carry out business at all. The futurist Alvin Toffler likes to ask executives what it would cost them in real cash if none of their employees had ever been toilet-trained. He is pointing out the enormous debt that corporations owe to communities, parents, networks and teachers.

How do they discharge this debt? Corporate responsibility, plain and simple. We have to rethink our approach to these issues as businesspeople. Our political postures must change. We have to stop endlessly whining for easier rules, lower costs and fewer restrictions. Our business practices must change too. We have to take longer-term views, invest in communities and build long-lasting markets. The new corporate responsibility is complex because it means changing our basic notions of what motivates us as businesspeople, of what our basic corporate goals should be. Unfortunately this is still a shocking idea for those people who believe it is a radical idea to consider anything other than financial profits.

When the share price was falling in June 1997, the *Sunday Telegraph* advised investors to get out of The Body Shop: 'Perhaps if the group spent more time pursuing a business strategy, rather than setting an environmental one, it would not be facing its current problems.' Although I don't want carpetbaggers or profiteers near The Body Shop, I believe the majority of our investors can see past such a narrow and mistaken view. The shares fell during that period, but it had nothing to do with how much effort we put into environmental and social causes. Our research shows that we gain very little in the short term out of being socially and environmentally responsible. We simply choose to be that way because we feel it is a more human way of doing business.

A SHORT HISTORY OF BUSINESS RESPONSIBILITY

The truth is that business as it is generally understood now, with its narrow concentration on profit, is not actually a very old idea. For most of history, excessive money-making has been consistently condemned as sinful and even after the Protestant Reformation changed society's attitude to business enterprise, corporations were chartered according to the duties they owed to the community as a whole. Queen Elizabeth I chartered the early trading corporations with this duty to the wider community central to their purpose. In many ways, social responsibility in business means reaching back to the original tradition that has been mislaid somewhere along the way.

The history of business, especially since then, has been peppered with the narrow wisdom of the robber barons and corporate giants that have shaped the myth of amoral business. John D. Rockefeller once boasted that he was quite willing to pay someone a salary of $1 million if he displayed certain brutal characteristics: 'He must be able to glide over every moral restraint with almost childlike disregard ... and has, besides other positive qualities, no scruples whatsoever and be ready to kill off thousands of victims – without a murmur.'

A whistle blower from The Limited sent the *Wall Street Journal* an extraordinary internal videotape called *WAR*, a 'motivational' tape on which rallying cries from top executives are interspersed with snippets of battle scenes from war movies where bombs explode and a weary soldier drags a wounded comrade through the forest. A parade of Limited executives speaks in mournful tones: 'You must be willing to shed fear and doubt.' 'Make the critical and key decisions effectively and efficiently, then move in for the kill.' A military marksman homes in on his prey and fires. 'This is not a mission for people who need developing or the weak-hearted,' the executives say.

Ruthless talk. But it demonstrates a myth that has clouded

business thinking for far too long. According to this myth, still rampant today, business and ethics don't mix. Business shouldn't be concerned with anything above or below the bottom line. Moralizing is out of place and any activity should be judged only as far as it affects the costs and benefits of 'good business'. I think we are paying for this pervasive myth, because it has shifted us from the idea of business as honourable exchange to one where cost-effectiveness has somehow to be balanced against justice. **The most significant clause in our trading charter, the charter that sets out the commitments of The Body Shop, is the promise to integrate principles with profits.** Conventional business may find this hard to grasp, but altruism also lifts the spirit, allows creativity to play freely. It gives other people the spotlight and keeps achievement in a healthy perspective. It is a wonderful immunization against narcissistic sickness. And I believe that once the way in which a human being is valued changes – once we are no longer judged by how much we are worth – then society will become more benign.

ETHICAL AUDITING

The question everyone should consider is this: can you act ethically if your company is listed on the stock market? Can you act in a really responsible way, for your employees, the community, your suppliers and customers, if you are part of an economic system that says you are measured primarily by your profit growth? Here the jury is still out, but the answer – if there is one – lies in how we measure the success of this kind of responsible business. If all we recognize is the narrow financial bottom line, then regeneration is going to be tough. But if we can audit the heart as well as the financial books, then there may be a way forward.

Auditing shouldn't just be for accountants, after all. If The Body Shop wants the freedom to campaign on public issues, we must first demonstrate our commitment to our beliefs. This means opening up to defined standards of human rights, social welfare and worker safety, environmental protection and, where relevant, wider ethical issues like animal protection. We believe we have a moral responsibility to tell the truth about ourselves and face up to those areas where we fall short.

That's why we have been trying to bring in new barometers of measurement that show that people and the Earth do matter. Every two years The Body Shop conducts a social audit, which offers a means of evaluation of the social impact and behaviour of an organization in relation to its stakeholders. We have 5,000 stakeholders – anyone involved in or affected by The Body Shop, from the suppliers to the staff to the shareholders to the communities in which we're working to people who have received money from the foundation – and the audit involves them all telling us what they think of the company. It's more than a staff survey; it's a full-scale in-depth assessment of our ethical behaviour. Our business is our stakeholders: our staff, our franchisees, our customers, our suppliers, our shareholders and the communities we serve. They tell us where we've done well, where we've screwed up and where we need to improve. In effect, our stakeholders provide us with a substantial blueprint for the future, because if we don't do right by them, then The Body Shop will be seen as no better than any old cosmetics company.

It is all very well undertaking social audits, but it isn't yet clear how to apply the knowledge and contributions provided by all those stakeholders for the benefit of the business and the communities in which it operates most productively. Until we do so, we still have to learn a great deal about how to communicate the new and heartfelt ethical information in ways that are useful and accessible.

OUR VALUES REPORT

'Ethical auditing' is an all-encompassing term that describes social and environmental auditing and any other ethics-related auditing that we may do, such as animal-protection auditing. We began it as an independently verified assessment of the company's performance against our stated values, and in 1995 we produced our first Values Report. The latter involved in-depth interviews and wide-scale surveys with all our stakeholders, ranging from employees to shareholders, from suppliers to local communities.

The fact that we were carrying out this enormous undertaking, and publishing the results, was showing that we were serious about transparency and accountability and about our commitment to these areas. It wasn't just a matter of publishing the stuff and then not thinking about it for two years, it was a public commitment to our values, a demonstration of what we were prepared to do to improve our performance – because we're not squeaky clean. It was also a warts-and-all study, telling people exactly how it was. 'As you can see,' we were telling people, 'it isn't a great glossy publication, it is incredibly boring to look at, but it has all of the information in here that anyone should ever want to know about our values.' It was no surprise to find we weren't perfect and that there were areas for improvement. And to make sure we did improve, we publicly committed ourselves to targets and to reporting our progress.

Values reporting is our contribution towards providing a measure of progress towards more sustainable operations. Our report is a public document, painfully honest at times, of our practices.

ENVIRONMENTAL AUDITING

Fifteen years ago, the idea of an environmental audit was popularly regarded as the provenance of the lunatic fringe. Now the principle has been internationally accepted, largely due to the power of public opinion.

The International Institute for Sustainable Development says that a meaningful sustainable development report should provide an honest accounting of the entity's performance – its successes and failures, strengths and weaknesses. It is not the responsibility of the corporation to value natural resources, it is society's responsibility. But in order to exercise that right, society must have information, so the responsibility of companies is to supply it. So we also produce an environmental statement, which is our contribution towards providing a measure of progress towards more sustainable operations. It is a public document and allows the flow of information that is so vital to creating a link with our stakeholders, who are very concerned about the future.

Having chosen to take an ethical position on the environment and to elevate conservation over consumption, a number of things follow. First, the ideals have to be shared by everyone connected with the business. That includes staff, subsidiaries, franchisees, suppliers and customers. Second, we are not just interested in good housekeeping, we also want political change: locally, nationally and internationally.

Saving 10 per cent on an electricity bill is not the most important thing. It is good to be cost-conscious, but playing our part in reducing carbon dioxide emissions and nuclear waste is much more exciting.

We place ever-increasing emphasis on renewable resources for our raw materials and infinitely recyclable components for our machinery and transportation. This will mean driving towards 'zero emission' of pollutants – not just for our own wastes, but those of our suppliers, and their suppliers too.

EMPOWERMENT AND EXPERIMENT

As already mentioned, I believe all life is an expression of a single spiritual unity. We can no longer afford false divisions between work and community, between ethics and economics. But how can we change from a system which values endless increasing profit and materialism to one in which the core values are community, caring for the environment, creating, growing things and personal development? Answer: we empower people. And in an organization, empowerment means that each staff member is responsible for creating that organization's culture.

One of the great myths of the 1960s is that rebelliousness belongs to the young and powerless; complacency to the old and powerful. We cannot afford to make that mistake again – we *all* have to create effective change. Everything we care about depends on it.

There aren't many motivating forces more potent than giving your staff an opportunity to exercise and express their idealism.

When I look at The Body Shop, it seems to me that there aren't any rules. No one has been there before. That gives us enormous freedom to experiment towards what we want. Believe me, it's a crazy, complicated journey. It really is experiment, experiment, experiment.

It is also an exhausting business breaking down barriers of traditional ways of working – it's a matter of trial and error, opportunism and sometimes quite literally 'Let's try lots of this stuff and see how it works.'

Socially responsive companies are often faced with more tyrannies and paradoxes than other companies: the tyranny of altruism, the tyranny of time, the tyranny of assumption and the tyranny of inadequate measurements and the tyranny of 'or'. But even with this list of tyrannies we choose to measure ourselves against a different standard, because we need to know we can make a difference. We want to measure our progress towards supporting communities and families, and to safeguarding the environment. If we get that, we have more chance of creating a business culture that encourages countries to educate their children, heal their sick, value the work of women and respect human rights. In short, to reclaim what was once an essential part of being human.

If business is managed from the heart, great things can happen. Which reminds me of a great slogan: 'Activism is the rent we pay for being on the planet.'

This place is run
like an SS
Concentration Camp

well you know what to do
if, you don't like it, don't you?

Anita for PM, OK?

Did you see Anita
on BBC2 last night?
Wasn't she wonderful?

What a lond of horse crap.

Anita is a maniac

PERSUASIVE PASSION

The lavatory is an excellent forum for communication. Everyone visits it at some time in the day and it is *the* great place to meet and chat with people from other departments, people you would not usually meet in the normal course of your work. At one point at our Littlehampton headquarters, we even formalized the toilet as a communications centre by inviting staff to leave behind their thoughts as graffiti. We hung a pencil and a pad of paper on a piece of string in every toilet cubicle, called it 'Throne Out' and it turned out to be an extremely effective way of getting uninhibited feedback. Here is a typical exchange:

'You can get as political as you like, but it should be remembered that this company is essentially a shampoo outfit. People come to work here mainly to get by in life – they should not need to support a pile of political pressures in their day-to-day humdrum work life. The politics should come out into the open and let the shampoo company make shampoo.'

'Form a pressure group independent of The Body Shop, but funded by The Body Shop.
That way neither the company nor the politics is degraded by the other.'

'I agree with him.'

'We work to live, not live to work.'

'True, but irrelevant in this context.'

Graffiti on the 'Throne Out' communications section in the toilets at The Body Shop headquarters, Littlehampton

'In my opinion this person is missing the whole point – i.e. that business and "politics" – if that's what you want to call action for change – are inherently inseparable. I for one sincerely believe this and say, "More power to our elbow." '

'I agree. If you don't want politics to interfere with your everyday humdrum working life, then you shouldn't be at The Body Shop. Try the civil service.'

'You silly, stupid bastards – what I meant was that the politics should be set up on their own so as to be able to act more strongly, independent of office apathy. That is honesty – where do you lot come from, Communist China? What a load of sycophants!'

I would guess that about 10 per cent of the graffiti was garbage, but a lot of it was really good and we had it printed up, along with replies from the board. I don't think the company made any important decisions as a result of the Throne Out messages, but it certainly reaffirmed our willingness to listen to the grass roots and from that point of view alone it was a valuable exercise. And it was a lot of fun as well.

Fun is important in communication, and so is passion. I believe that communication is the most important tool of leadership and passion is the most important element of communication. It is passion, above all, that persuades. For all its modern emphasis on communication, so much of business forgets this crucial element.

When I was young, I saw a headline in a newspaper that touched a nerve in me and some 45 years later it is still resonating. It was the front page of the *Daily Mirror*, then the most respected campaigning tabloid newspaper in the UK, at the time of the Soviet Union's defence of the Hungarian invasion at the United Nations. The headline was in the boldest type imaginable and was spread over the entire front and back page. It read: 'Mr Khrushchev, don't be so bloody rude!' And underneath in a smaller box it said, as a kind of afterthought: 'Who do you think you are? Stalin?' It was a magnificently outrageous way to get the message across and from that day onwards I understood the power of communicating with passion.

Leaders have to communicate in ways that move people to

The passionate are the only advocates who always persuade. The simplest man with passion will be more persuasive than the most eloquent without. René Descartes action. But competence must be blended with compassion if a leader is going to be effective, because it doesn't matter how much you care, if you can't *communicate* that care, you might as well not be there.

Communication is the key for any global business.

As a business, you have to listen to your customers at exactly the same time as you are telling them where you're coming from. And you have to keep reinventing the communications arm of your business while still remembering why you started out in the first place. It is a big mistake for any company to separate itself from its history, even one only 10 or 20 years old. In many cases, retailers lose impetus when they forget what inspired them to start up.

As we rush to gain a global perspective, business leaders also need to stop and reflect not just upon why they went into business, but upon what continues to make them different now. Are they growing for growing's sake or do they have something more to achieve? More importantly, do they have something better to offer? If so, is that clear to the marketplace? Are they talking to their customers?

I am constantly confronted and confounded by myths in the world of business. One of the biggest myths is that the more you expand, the harder it is to communicate. This is clearly true sometimes, but I haven't found it to be *necessarily* true.

The company I founded in 1976 now trades in over 2000 shops, in 51 markets, across 12 time zones and in 24 languages. It has to keep franchisees and employees updated and motivated while driving forward with new concepts, new store designs, new products and new ideas for communication. Communication is vital.

I used to wish that by the millennium we wouldn't be a cosmetics company with a communications arm, we would be a communications company with a cosmetics arm. There was a time when we annually produced an avalanche of posters, pamphlets, magazines, leaflets, flyers, stickers, videos and papers on every subject

under the sun, from the inevitability of the ageing process to the iniq-
uities of the regime in Nigeria. We still bombard our staff with information, our communications network includes bulletins that go out to
all staff to keep them up to date with new ideas, new developments
and new products. Staff are family at The Body Shop: we encourage
debate, encourage employees to speak out and state their views. I'll
never forget a long-standing employee saying to me once: 'I've never felt so much part of a family in any job, because everyone is
involved in everything we do.' I stick by that.

THE USELESSNESS OF ADVERTISING

We are assaulted by tens of thousands of advertisements every
year, women more than men, and the din of advertising has grown
so loud it is impossible to tell one pitch from another. So I believe
that conventional marketing techniques are increasingly ineffec-
tive. Customers are hyped out. They have been over-marketed. They
are becoming more cynical about the whole advertising and marketing process. We have learned that to educate and communicate
in the contemporary business climate you have to be daring,
enlivening, different and willing to take risks. We have learned that
marketing is about being able to communicate more sensitively and
more persuasively with the consumer.

We're in the skin and haircare business, and it's an industry
not known for treating women with respect, whether it is targeting
women with anti-ageing creams, unnecessarily testing products on
animals or wrapping products in wasteful and unrecyclable packaging. People know what is going on and they increasingly want to
act on the information they have. Unfortunately, the facts they need
to make an informed decision are often obscured, confused or
unavailable. Customers crave knowledge and they want honest
information. We need to get the message across.

The Body Shop began campaigning in order to capture and
channel the imagination and outrage of individuals who, collectively,
can do something about corporate misinformation. We use many

different ways to do this – everything from our shop windows to billboards, the sides of our lorries and staff uniforms. But we rarely advertise in the traditional sense. We provide information, we make everything clear – from our ingredients to our values – and let customers make informed decisions about whether they want to do business with us.

It may seem unfathomable to many people that The Body Shop can sell cosmetics without mass advertising, especially in a business where even the department stores routinely spend millions of dollars on glitzy advertising campaigns. But we do not believe in revving up demand for a product by throwing huge amounts of money into advertising. It has been described as sewing without a needle or preaching without a pulpit – whatever, the analogy doesn't matter. Some of our products have stories – their history, how they were con-

The growth of The Body Shop is testimony to the fact that you don't need to waste money on costly advertising campaigns to be successful. Instead, we've always relied on word of mouth and stories.

ceived, where the ingredients come from, how they are made and anything else people might want to know. What are you most likely to remember – some ludicrous advertising claim that you know to be a lie or the fact that the cream you are buying comes from nuts gathered by a women's co-operative in Africa? So we sell cosmetics with the minimum of hype and packaging, and promote health rather than glamour and reality rather than the so-called 'promise of eternal youth'.

GUERRILLA MARKETING

I like to describe what we do as 'guerrilla marketing'. That means using unconventional, low-cost tactics to get attention. We turn our shops into action stations for human rights, for example. We lobby our customers to speak out on issues that affect them.

I hate blank spaces: I view an empty space as an opportunity to create an atmosphere, deliver a message and make a point. What could be more boring than a blank paper bag? Put a message on it!

I believe in promoting our products by linking them to a political and social message.

The Body Shop has a fleet of 12 lorries that travel the length and breadth of the UK delivering products to the shops. They act like moving billboards, carrying out onto the highways messages that inform, educate or entertain the public. I'm not talking about advertising a logo here, I'm talking about things that really matter – like missing persons. The first time we did it, we painted the faces of four missing persons onto our trucks, along with the names of at

least a dozen others who had disappeared and the number of a telephone helpline. More than 30,000 calls were made to the helpline and some of those people were found. A girl who had been missing for more than two years was discovered in Marseilles because her face was printed on one of our trucks. That's how you connect with the community.

These initiatives have a dramatic effect because they make people feel they are part of the great concerned family of humanity and make them feel they can make a contribution, no matter how small. We are making it easy for people to respond and to get involved. Giving them a phone number to support the efforts of the Missing Persons Helpline will not increase sales of a specific product, but it does create a banner of values. It links us to the community and

makes us feel included.

One of our projects in India has been described by the World Health Organization as one of the most effective grassroots AIDS education campaigns it has seen. We use an elephant. The elephant walks down the streets of towns in Tirumangalam in Tamil Nadu with information on its side in Tamil and English about where to go for information on sexually-transmitted diseases and how to get free condoms.

In another part of India we run a 'direct action' contraception campaign. When truck drivers stop to refuel at one of the major stops on the main road going up to Madras, they pay the waiting prostitutes, go over the wall and do the business. Our workers, most of whom are young medical students, hotfoot it over the wall, tap the customers on the shoulder at the appropriate time and offer them free condoms.

I am not necessarily suggesting this is a campaign that can be adopted by everyone, but I am suggesting that there is business advantage in doing something different, rather than apeing the competition. **And I firmly believe that taking a high profile in the community is a far better marketing strategy than trying to outspend the competition on advertising, if for no other reason than the positive impact on staff morale.**

LEADERSHIP BY STORY-TELLING

The people I work with are searching for something more than just doing a job – they also want to learn and find meaning in their life. They are open to leadership that has a vision, but this vision has to be communicated clearly and persuasively, and always, always with passion. And communication is not just about verbal or visual language, it's about body language too – how you hug and embrace members of your staff, for example.

For a leader to motivate and gain trust and loyalty, a message via e-mail is not enough. Suggestion boxes are not enough, either. Nor are employee surveys. Nor are graffiti facilities in the toilet. Nothing, but nothing, can substitute for a live leader, listening and responding.

Despite the wonders of modern communication, the leader with a serious message has a problem. Today's is a big noisy society. The leader has to compete with popular entertainment, skilful advertisements and the drama of current events. I believe that one of the most effective means of communication is story-telling. Stories are humankind's oldest way of communicating and comprehending truths. Most of us were probably brought up on fables or fairy tales, either at school or at home, and story-telling is an excellent way to learn because of that.

Stories 'excite the imagination of the listener and create states of puzzlement, insight and resolution', according to one researcher in speech communication. 'The listener is not a passive receiver of the information but is triggered into a state of active thinking.' I know this is true from the time I have spent with tribal peoples, who often invest story-telling with mystical powers.

My own life as a child was dominated by story-telling. There were no children's rhymes or riddles for us. Instead, my mother told us stories of romantic love and deep feeling, stories of life on her farm in southern Italy, stories of how we were conceived and how we were supposed to relate to each other. We heard family anecdotes galore and tales about irreverence towards the Church. There was more majesty in the stories my mother told than in any organized religion.

I still see story-telling as a major component of communication within The Body Shop, both stories about products and stories about the organization. Stories about how and where we find ingredients bring meaning to our essentially meaningless products, while stories about the company bind and preserve our history and our sense of common purpose.

I read somewhere that in every company there is a hidden network of communicators who are important because they not only disperse information, but also interpret it. Dividing this network into story-tellers, priests, whisperers gossips and spies:

- ☞ *Story-tellers* preserve the spirit and meaning of stories told about an organization and impart these legends to new employees. As in tribal societies, a company's stories serve to create and recreate what is meaningful for the employee.
- ☞ *Priests* are the custodians of the spirit of a company, its image and values. A healthy workplace is one where the values are discussed. Values that are never expressed tend to be taken for granted and inadequately conveyed to newcomers.
- ☞ *Whisperers* are those who disseminate information covertly.
- ☞ *Gossips* know everything, but are never expected to get the news right and are never taken seriously.
- ☞ *Spies* seem to me to be something every good senior manager ought to have to keep them informed about what is going on.

We realized that we needed to learn more from our own story-tellers within the company because the penalty for failing to listen to stories is to lose our history and the values we strive to promote. As we have grown, the stories that have been told and retold about The Body Shop have entered the chronicles of the company, but our growth has had an inevitable consequence – the loss of good old-fashioned belly-to-belly conversations. We try to counter this by generating an ongoing dialogue, world-wide, between the company and the staff, shops and customers, business and community.

COMMUNICATING BEYOND HIERARCHIES

Almost all large corporations are run like military institutions: there is strict hierarchy. No one from the lower echelons gets to talk to anyone at the top and God help you if you park in the chairman's space. We wanted to prevent a similar hierarchy developing in The Body Shop, so we instituted a 'red letter' system to open up communication through all levels of the company. Any member of staff, anywhere in the world, who is unhappy, or bugged about something, or having problems, or just wanting to get something off their chest, can put their thoughts down on paper, stick it in a red envelope and be sure that a member of the board, will see it and respond within 24 hours.

In our offices we have no areas off limits to anyone – cleaners are welcome to use our executive boardroom for meetings. In earlier days, we encouraged late night mind-blowing conversations with staff, the kind none of us had had since college, breaking down barriers between boss and employee, abandoning taboos that limit who can talk to whom about what. We ran family days and company days to help familiarize staff not only with their co-workers but also with activities in other areas of the company.

Gordon and I are very informal people. We would have been climbing the walls in search of an escape route long ago if we had to operate in a stifling, traditional business culture.

The key to handling problems and conflict within an organization is to keep the channels of communication wide open.

Among other experiments, we held a series of lunchtime lectures on subjects ranging from whether The Body Shop should be unionized to creation theology. At one point, we even had a violinist offering music therapy to staff.

EYES WIDE OPEN

One of our most successful communication tools was in-house video. Most of our staff are under 40 and were raised on sound and vision

The only thing that matters
is how you touch people.
Have I given anyone insight?
That's what I want to have
done. Insight lasts; theories
don't.

Peter Drucker

business as unusual

bites. They expect to be communicated to in this way. All companies know this, yet they usually produce videos that are as dull as ditchwater, completely unengaging and therefore defeating their purpose.

For 10 years we produced a weekly programme for UK stores and a monthly one, translated into 21 languages, for stores worldwide. As far as I know, it was the biggest investment in business video in Britain. Video is good at motivating people, unlike print, which is fine for information but not much else. Video helps get messages and attitudes across to staff around the world; everyone understands a crowd marching behind a banner and demanding change. For 10 years our videos motivated thousands of employees around the world and it was to my real horror that the scheme was pulled back because it was seen not as an investment in the spirit of the company but as an expense.

But I have always tried to bring beauty, style and excitement into the office. I like the concept of a workplace that is as visually stimulating as you can make it. We hung art not just in the boardroom, but everywhere. All over The Body Shop the walls were hung with photographs, blown-up quotes, charts and illustrations. We had eight life-size men flying about near the ceiling and 12-foot long bottles and bananas in our warehouse. There is still a full-scale fibreglass replica of Manet's *Déjeuner sur l'Herbe* in the gardens outside the offices in Littlehampton. It's funny and it's charming. There was a whimsical dummy of a cleaning lady with a mop and bucket in our entrance lobby. We had Oscar Wilde quotes painted in two-foot high letters on walls and exhibitions about human rights issues in our hallways.

The language of design is very important to The Body Shop. Our messages tend to be visual rather than verbal and they are very rarely word-dense. It is really vital that we create a style that becomes a culture. It may be designed, but whatever we do, we have to preserve that sense of being different and of doing something that hasn't been done before.

We believe shop windows are one of the most important factors in promoting the image of the company. I think they should be provocative and edgy. We cannot afford to have our message diluted or confused by shop staff who execute a window poorly, or by new brand managers or designers who want us to appear less of a 'visual rabble rouser' and more of a 'contemporary sophisticated cosmetics company'.

INVOLVING CUSTOMERS

Of course, none of this is of any use if you're not communicating with your customers. I think there are few companies in the world that can claim to enjoy the strong, enduring relationship with customers that we do. We bring our customers into the heart of the company, we

think of them as family, we invite and encourage them to participate in everything we do. None of our competitors have been able to match us in this.

But our single most effective way of communicating with customers, after years of suggestion boxes or focus groups, is The Body Shop at Home. Some 16,000 consultants from the UK, Canada, USA and Australia, hold on average 250,000 parties with more than 2.5 million customers each year and bring back anecdotes, spontaneous reactions and stories. Nothing, but nothing in the history of this company has been more profound or focused as that two-way conversation in people's homes in terms of letting us know whether or not we are living up to our customers' expectations.

'STOP THE WAR!'

I make no apologies for driving the communications of The Body Shop, but I have to take responsibility for them, too, and at one point that responsibility could have cost me the company. At the very least it endangered my position.

I was in Boston when the Gulf War broke out and I remember being deeply impressed by how quickly the unions and churches there got billboards out calling for an end to the conflict. There were huge posters everywhere just saying: 'Stop the War!' As a committed pacifist, it moved me to the quick. I have been anti-war ever since my teens and the Gulf War seemed to me to be nothing more than a shooting arcade. **I was horrified by the idea of going to war, especially on behalf of a country known to be repressive – and for no more than the commercial implications of oil production in Kuwait.** When I got back to Littlehampton, I mobilized all the resources I could muster to put together an instant anti-war campaign. We got hold of every billboard we could lay our hands on to display quotes on the subject of peace and the futility of war, quotes from everyone from Jesus Christ and Mahatma Gandhi to Winston Churchill. I put them up without permission. We even posted the War Resisters' League line, which said: 'It will be a great day when our schools get all the money they need and the air force has to hold a jumble sale to buy a bomber ... ' We set up petitions in each of the shops to protest against the war and encouraged customers to send faxes to President Bush and Saddam Hussein calling for the conflict to be stopped.

it will be a great day when our schools get all the money they need and the air force has to hold a jumble sale to buy a bomber...

There was huge enthusiasm in the British media about going to war and, almost for the first time, I felt very proprietorial and territorial about what we were doing. I had always championed the cause of peace and been against violence in any form, and it seemed to me that, on this issue, the values of the company were non-negotiable. When one of the board members rang me up in

America, telling me we had to take the posters down, I said: 'Over my dead body.'

The town of Littlehampton was confused by our stand. Littlehampton is a blue-collar town. Most of the grandparents have probably experienced war, but the kids have no notion of it, and what I was doing seemed to oppose the prevailing mood. It also split the company right down the middle.

About a week after the campaign had been launched, I went back to the United States and while I was there I heard that there had been a board meeting and the decision had been taken to abandon the campaign and to remove all the billboards and the petitions from the shops. I was furious. For me it was a metaphor of what was wrong with business. Conventional business thinking decrees that someone like me should not be in a powerful position in an organization whose main purpose is to make money, because I 'mush everything up'. I act as if I am working for a non-profit-making organization. I campaign on issues that have absolutely nothing to do with the cosmetics business. But my strength is that I don't give a damn about conventional business thinking. If we ended up like every other large corporation, ignoring our values and history, I'd die of boredom, as would hundreds in our company.

The prime mover behind the decision to stop the campaign was our then UK managing director. **He believed it could damage the company's image and marketability. I believed we were morally obliged to speak out.** He had lobbied the other members of the board and persuaded them that we should not be getting involved in issues like the Gulf War. I think it was his stab at trying to make The Body Shop a more 'normal' company. I wasn't about to allow it to happen.

I needed to be able to convince everyone that the stand I was taking was right and it seemed to me there was only one solution – a full-scale debate involving all the staff at head office. Actually very little public debate had taken place – the media was obsessed with the story and everyone seemed to be war crazy. I felt very alone. So we closed down the company for a day, set up a video link with the London office and hunkered down to discuss whether or not The Body Shop should be campaigning for an end to the Gulf War.

I had my heart in my mouth. What if the whole company said: 'Anita, we must stop this – we have to follow government policy'? I knew if that was the democratic decision I would have to follow it, but I felt willing to leave the company on this issue of principle. I felt I had no option.

But I also had a card up my sleeve. A few months earlier I had taken a cross-section of staff – everyone from senior managers to fork-lift drivers – up to my house in Scotland for a few days to talk about the company – what it was, what we would like it to be and what its vision should be. We had a great time, cooked together, walked in the fields a lot and talked and talked. One day, two of the guys in the group mentioned that they had been in the army – one had served in Northern Ireland and the other in the Falklands – and they began describing, very quietly and movingly, the reality of actually fighting in a war, the sheer horror and dread of being in battle.

I knew if I stood up at the meeting and proselytized that I wouldn't be half as effective as them, even though I think communicating is one of my strengths. So I let them do most of the talking. In front of 500 employees, they spoke from the heart about their experience of war and how it had altered their lives. They talked about fear and squalor of war, about watching their friends die, about the sheer horror of the experience. They did all the work for me – and they saved me. When it came to taking the vote I feared that my fate was in their hands, but I needn't have worried. The vote was overwhelmingly in favour of continuing with the campaign. I drew an enormous sigh of relief and felt a part of my company again. I knew I had been close to the brink.

I have often sat and scenario-played about what might have happened if the vote had gone the other way, if they had said: 'Oh come on, Anita, get real, it's time we stopped this nonsense.' The worst possible outcome would have been total estrangement from the company and all it stood for. The best, I think, would have been to have left honourably, saying: 'This company no longer stands for the things I stand for.' Somehow – and I have played the scene time and time again – I always come round to thinking that in the end, despite everything, I would have stayed. When I analyse why, I think it is simply because I had nowhere else to go. And it would have been like leaving a child – I had given birth to it, shaped it all those years and I still hadn't finished with all the things I wanted to do with it.

As a result of this episode we decided that the company's core values should at least be documented in black and white, even if they weren't actually engraved in stone, and we put together The Body Shop Charter. This took 18 months to complete, but it became a strong symbol of grassroots participation in management and established the 'red letter' procedure. The first statement clearly demonstrates our commitment to integrate principles and profits: 'The goals and values of The Body Shop are as important as our products and our profits.' These were to be more than just empty words: eight working groups were set up to institutionalize the values and make sure they penetrated every nook and cranny of the company's operations.

THE BIG ISSUE

Of all our communications initiatives, I think I am probably most proud of one that really does not involve The Body Shop at all, at least not now. In July 1990, Gordon was in New York and he bought a copy of *Street News*, a magazine sold on the streets of the city by the homeless for the benefit of the homeless. Gordon thought it was a brilliant idea, he was impressed both by the quality of the magazine and the guy who sold it to him, and when he got home he asked his friend John Bird to carry out a feasibility study to see if it might be possible to start a similar newspaper in London.

John is a very old friend, an off-the-wall Marxist. He went out and talked to homeless people in London, asking them what they would think about selling a newspaper on the streets. The

Honesty is the best image.

Ziggy (Tom Wilson)

message that came back was that anything was better than begging. When John reported to Gordon that he thought the project might work, Gordon presented the case to the trustees of The Body Shop Foundation, eventually got them to agree to put up seed capital and *The Big Issue* was born with John Bird in the editor's chair.

It was a big risk. No one had ever tried to involve the homeless and the disadvantaged in something like this. It could have gone badly wrong and there was always the danger that we would be accused of exploitation, or promoting drug abuse, violence, alcoholism, welfare fraud and heaven knows what else.

None of these things came to pass. After a shaky start, *The Big Issue* was soon judged to be a success, largely because it was a great read, was well designed and was not just a guilt purchase. Within two years of its launch it had gone from monthly to fortnightly to weekly, around £1.5 million had been raised for the homeless and some 1,000 homeless people had been registered as vendors. Selling *The Big Issue* helped put the vendors back on level ground and restored their dignity. With work has come a sense of security and with security has come freedom. Now some of the original street vendors have full-time jobs working for the magazine or for one of the innumerable training and counselling programmes it has spawned offering advice on finding employment, permanent accommodation and help with drug and alcohol problems.

We'd gone in with head, heart and body to create an effective partnership with *The Big Issue* and it was music to our ears when we learned in 1995 that it was setting up its own foundation to develop the creative skills of the vendors and offer training and education, housing resettlement and outreach programmes.

The Big Issue is now the fastest growing magazine in the country – which is absolutely brilliant – and editions are sold on the streets of cities across the United Kingdom. Satellite editions are produced in Namibia, Cape Town, Melbourne and Los Angeles, where it is distributed through bookshops and coffee shops.

While we provided the initial funding – The Body Shop contributed around £500,000 to get the magazine started and to see it through its launch – it didn't do anything for the company and we never expected it to. But I believe it set an example to other businesses.

Many governments' economic agendas seem to take no account of caring for the weak and the frail and the marginalized. If governments are not interested, then I believe that business – rich, powerful and creative – has to take responsibility. If not us, who?

FULL VOICES

In the late 90s we developed our communications strategy into a full-scale publishing operation covering a whole range of single-issue topics of interest to The Body Shop customers, not just health and beauty. Its genesis was a series of little booklets we produced under the imprint Full Voice. The first Full Voice was originally intended as a single-issue pamphlet to familiarize staff with our commitment to raising awareness about self-esteem. The message was: 'Like yourself the way you are.'

I had read in America that anorexia and bulimia were increasing among very young women, some as young as 12, and that every time they picked up a magazine and saw page after page of super-thin supermodels, their self-esteem plummeted and their condition worsened.

I thought it was shocking and the kind of problem that would strike chords among our customers, so in 1997 we produced the first Full Voice, a 28-page booklet with the title *The Body and Self-Esteem*. The message was encapsulated in a centre

spread featuring Ruby, a blow-up model with proportions more like most women, and the caption: 'There are 3 billion women who don't look like supermodels and only 8 who do.' It included a mass of facts and figures emphasizing the importance of physical and cultural diversity and the difference between the ideal and the real. We pointed out that 20 years ago models weighed only 8 per cent less than the average woman, whereas today they weigh 23 per cent less.

I was very pleased with the first issue of Full Voice and very disappointed that it appeared to raise so little interest when we put it in the shops. In desperation we decided to distribute it with the *Independent on Sunday* newspaper – and the response was mayhem. Everyone from women's groups to prisons to Girl Guides was suddenly clamouring for copies. Overnight, it seemed, it focused such attention on the issue of female self-esteem that everyone was talking about it. Magazines and columnists and politicians took up the subject and a number of 'self-esteem in the community' pro-

A psychological study carried out in 1995 found 70 per cent of women experienced feelings of depression, guilt and shame when they looked at photographs of models in magazines.

business as unusual

grammes were launched. In the United States, the booklet was inserted into *Lilith*, a Jewish feminist magazine with a big media readership, and so it generated a great deal of attention over there as well. We eventually sent copies to almost every country in the world and it is still being talked about.

It was so successful that we decided to turn Full Voice into a series. The next issue was about political activism and we followed up with issues devoted to the dangers of globalization, the benefits of hemp, the need to develop community trade and the horrors of sweatshop labour in the Third World. But then, as I said, passion is the key to communications, passion plus learning from your mistakes and not taking yourself too seriously.

Full Voice is one step towards that day when we'll be able to love ourselves inside out. Message sent by The Body Shop in Pleasanton, USA

Leadership is fundamentally about communication and dialogue, but it is also about having a dream and a vision and being able to develop a shared sense of destiny, showing others how they can realize their own hopes and desires within that vision. At the same time, I don't want The Body Shop to ever lose its sense of fun: passion, fun and vision – communication needs all three.

THIRTY-FIVE-ODD YEARS AFTER
THE WOMEN'S MOVEMENT
STARTED, WOMEN'S MAGAZINES AND THE
BEAUTY BUSINESS ARE STILL
FULL OF ADS BOMBARDING US
WITH IMAGES OF SMILING AND PASSIVE
WOMEN WHOSE MESSAGE SEEMS TO BE:

'SHUT UP, GET A FACE-LIFT
AND STOP EATING.'

THE TYRANNY OF THE BEAUTY BUSINESS

Ruby, the generously proportioned doll who first appeared in Full Voice in 1997, also filled the windows of the branches of The Body Shop that year. She has since appeared in magazines, newspapers and television all over Europe and gone on to take Australia, Asia and the US by storm.

Ruby was a fun idea, but she had a serious message. She was created by The Body Shop to challenge stereotypes of beauty and counter the pervasive influence of the cosmetics industry, and in doing so she kick-started a world-wide debate about self-esteem and body image. Ruby was here to say, 'If you **feel** gorgeous you'll **look** gorgeous.'

Looking at Ruby was supposed to remind people that beauty is about confidence, rather than the circumference of your thighs.

Ruby was not without her detractors. In the US, the toy company Mattel threatened to sue us because they claimed Ruby was denigrating the image of Barbie, their twig-like bestseller. I was thrilled by the idea that Ruby was insulting to Barbie – the

Notes for my article in the *Daily Mail*, February 2000

There are 3 billion women who don't look like supermodels and only 8 who do.

 THE BODY SHOP
KNOW YOUR MIND LOVE YOUR BODY

notion of one inanimate doll insulting another was absolutely mind-blowing. Then in Hong Kong, posters of Ruby were banned on the Mass Transit Railway because the authorities said she would 'offend' passengers. Of course, the much more seriously offensive pictures of pneumatically-breasted blondes remained on the railway for all to see.

And there is my relationship with the beauty industry in a nutshell. It makes me angry – not just because it's an industry dominated by men trying to create needs that don't exist, but for what it does at its worst. At its most extreme, the beauty industry seems to have decided it needs to make women unhappy with what they look like. It plays on insecurities and self-doubt about image and ageing by projecting impossible ideals of youth and beauty. It blinds us with science without giving us the kind of practical information we could use. And it has rarely celebrated women outside Caucasian culture. But then I don't believe an industry which so many women find so unsettling could really claim to celebrate or cherish women of **any** culture.

Leonard Lauder, son of Estée, once refused to advertise in Ms magazine because, he said, his products were for 'the kept woman mentality'. What a bizarre signal to send his female customers. I'm sure he lived to regret his declaration, but the bad taste still lingers. To this day, it crystallizes my suspicions about the business of beauty.

MAGNIFICENT BODIES...

One of the beauty industry's biggest lies has always been that you can turn back the clock with a face cream. Even if this defies common sense, millions of women – and men too – have been suckered by it. After all, who wouldn't want to believe in something as easy as hope in a jar? There is no cream in the world that will restore youth to a 50-year-old woman. But for some reason, we let the beauty industry sell us that hope. For some reason, we let it portray female flesh as gross and in need of repair. For some reason, we allow it to portray youth as the ideal. We accept that the industry will spend millions of pounds concocting dubious potions to 'cure' conditions like cellulite and ageing – natural processes that occur in all women – and yet fail to broadcast the *real* messages about health. **Why do beauty magazines devote thousands of column inches to cellulite rather than exposing the tobacco industry's efforts to seduce a female audience?**

...BEAUTIFUL FLESH

The fact that so many women are unhappy with their bodies these days is evidence that the strategy of the beauty industry has worked. Something is badly awry in our world when half of all 11-year-old girls are dissatisfied with their bodies, when symptoms of anorexia have been detected in eight-and nine-year-olds and when you have idiots in the media like the Manhattan doctor who joked: 'It's much more important to be thin than alive.' The point isn't so much that the beauty business is directly responsible for the eating disorder epidemic – it's not as straight forward as that – but it doesn't seem to recognize its responsibilities to its customers.

Statistics tell a disturbing tale. The US diet industry is worth $77 billion a year. Over half the female population of the US buy into it and a survey of 10-year-old girls found that 80 per cent were already on diets. Yet as many as 98 per cent of all those using the products and services supplied by this industry fail to achieve their objectives. In other words, the products and services don't work. The diet industry, America's fifth largest, is probably one of the most successful marketing achievements in history. What it sells is self-doubt, and it has relentlessly and successfully extended its grip on the minds and bodies of millions of women all over the world.

...BEAUTIFUL WRINKLES

The universal preoccupation with youth and conventional glamour creates a callous society in which women diminish in status as they age. Between them, the media and the beauty industry have successfully alienated us from our own bodies – and our own lives. Stretch marks and wrinkles show how we've worked hard, in and out of the home, raised kids, enjoyed good meals, tossed back a drink or two, laughed, cried and struggled. They are just symptoms of what gives our lives value. And yet the *wisdom* we have acquired is of no value compared to our appearance.

I think it is my job, and the job of all women my age, to redefine the notion of beauty and to legitimize the ageing process. We have to spread a new message.

We have to tell ourselves: 'If you really didn't ever want to get wrinkles, then you should have stopped smiling years ago!'

In my ideal world, the way we age would be a cause for celebration, because it is one of the elements that defines us as human beings.

If our culture chooses a cult of eternal youth, it will actually be something of an own goal, as almost 25 per cent of the UK's population will be over 65 by the year 2025. Women aged 65 and over are already the fastest growing segment of the British population. It seems to me that instead of building up prejudices against ageing, we should be looking for the silver lining – or a Golden Plan, which is the name of the social and medical research programme launched by the Japanese to make fundamental changes in their society so that longer lives will also be more fulfilling ones.

CREATING UNFULFILLABLE DESIRES

Today we are subjected to a constant barrage of idealizing claptrap from the cosmetics industry, assaulted by 30,000 advertisements a day. Most are targeted at women. And when any newspaper, any magazine, any billboard is directed towards women, it's about control. If you can control the shape of women, you can control their thinking as well. On every level, the cosmetics industry is about perfection, but the subliminal message is: 'We can control what you look like.' When we look like ourselves, we all look different, but under the control of the beauty industry we all start looking the same. Every nose job, every facelift looks alike; they seem to inspire a need for homogeneity and for uniform looks. Where's the celebration in that?

Advertising beauty products is so easy. Here's a step-by-step guide:

1 Create a lotion that feels good on the skin. (The fact that it really doesn't do much good doesn't matter – all you have to do is imply that it works wonders.)
2 Take a photograph of a high-cheekboned, slender, athletic-looking woman and retouch the photograph to get rid of wrinkles, cellulite and anything else you don't like.
3 Put your product next to the picture and let the public make the obvious assumption that one follows from the other.
4 Dismiss claims that you are making women feel inferior and unhappy with how they look as hysterical. Dismiss the connection between advertising and the mental health of young women as the bizarre ranting of feminist lunatics.

HAVING THE GUTS TO DO SOMETHING DIFFERENT

We don't approach the issue of beauty at The Body Shop as an élitist preoccupation with passive perfection. **For us, beauty is a healthy part of everyday life. It's all about character and curiosity and imagination and humour – in short, it's an active, outward expression of everything you like about yourself.** What we have consistently done is to try to go in the opposite direction to the industry. We celebrate women rather than idealize them. We sell cosmetics with the minimum of hype and packaging and promote health rather than glamour, reality rather than the dubious promise of instant rejuvenation. Right from the start, we wanted to be honest about the products we sold and the benefits they promised. We will not be tyrannized into going along with the rest of the industry with the usual false blandishments.

We do what we can to challenge the concept of femininity as portrayed by the beauty industry and we work instead to promote self-esteem, cultural and physical diversity, and encourage the celebration of the unique qualities that make each of us what we are. If the beauty industry makes us celebrate products supposed to change what women look like, rather than celebrating what women have done, then society becomes shallow. There's such a poverty of praise in our society towards women already: they are not taught they are remarkable. And historically, remarkable women have often come to sticky ends. Under the Inquisition, three million women were burned at the stake because they were herbalists or midwives or simply wise. It's a cultural history that the cosmetics industry is doing nothing to change.

The beauty business tends to describe products as if they were the body and blood of Jesus Christ. And they try to legitimize their tactics by marketing from a pseudo-medical platform. They have to pretend to be a branch of the pharmaceutical industry. They don't copy us by saying that any old moisturizer works. Nor do they follow us in encouraging people to use a product simply because it has fun ingredients or because it's sourced from the Third World. Instead, they offer up pseudo-medical evidence for outrageous and ridiculous claims.

I'm tired of all this nonsense about beauty being only skin deep. That's deep enough. What do you want – an adorable pancreas?

Jean Kerr

No matter what chemical formula is on the label, no cream will make your breasts bigger or your thighs thinner. No shampoo will get rid of split ends, no matter what the manufacturers claim. If you want to get rid of split ends, cut your hair. All a shampoo will do is clean your hair. That's it.

People spend £10 billion a year on toiletries and cosmetics around the world. A significant amount of that is spent on moisturizers and cleansers. A moisturizer is really the only valid skin-care product, because it really does help prevent loss of moisture in the skin. But it won't make you look younger. We tell our customers that. Does this fact help sell our moisturizers? Probably not. Does it go some way to put wrinkles in perspective, to give people another view on the ageing process? I hope so.

We go even further in our shops in California, encouraging people to embrace and celebrate their wrinkles. Most of what we think of as ageing is actually exposure to sunlight, so the most effective anti-ageing product is a sun hat. **This is what we mean when we say we have a greater responsibility to our customers than simply tending to their looks. We try to celebrate the whole person.**

HONOURING THE INSTINCT

When I spend time with women in other parts of the world, I find the most direct route to their hearts is always through stories of common rituals, both the big ones – birth, marriage and death – and the smaller intimacies of cleansing and caring for their bodies and hair. Women in Polynesia, women in Africa, women in

Sri Lanka all use skin and haircare as an excuse for a get-together, a chance to sit around and share stories, because women are born story-tellers. In some cultures, that is their main value as they age.

These kinds of rituals are one stage on a journey which challenges almost every notion of beauty as we see it in the West. It acknowledges that our survival as a race lies in recognizing and accepting diversity – cultural, biological and any other way you choose to label it. So the rituals of beauty are actually more important than the products – they offer a focus, a way to express self-respect and a means of social bonding.

You can find similar impulses in our own culture. When I was young, we would borrow each other's mascara and do each other's hair, and this would lead naturally to the intimacy of shared stories and experiences. Everything – from choosing a product to the ritual of application to the appreciation of the finished product – was calculated to make you feel as though you were doing something special for yourself. The results were that you would look good and you would feel better about yourself. It's powerful stuff, and it pisses me off that the beauty industry has abused that power.

THE EYE OF THE BEHOLDER

Curiously enough, it is only in the last two centuries that appearance has defined beauty.

The idea that beauty is nothing more than a combination of features is ludicrous: it's about action, vivaciousness, courage, energy and compassion – all the things that women should be celebrated for. It's not passive, not a combination of high cheekbones and bee-stung lips.

This is a very easy point to come to grips with if you take a look at what is considered beautiful around the world. Beauty is by no means universal. For instance, the Maoris admire a fat vulva and the Padung go crazy for droopy breasts. In Kenya, I found that Swahili men go nuts for nice hands and feet – just

about all they can see because their women are veiled. The Ashanti consider women who would be considered downright fat in our culture to be the most alluring. The same can be said of many West African cultures, where women are appreciated as beautiful if they are heavy-set, with large breasts and a large bottom. And it is the accumulated wisdom of fairly advanced age that a young Aboriginal man finds attractive in women.

I think the fashion and beauty industries miss every opportunity that is offered them to celebrate such diversity. They're in the business of reducing hopes and aspirations to easily marketable formulae, so they tend to address surface or superficial concerns. They need to establish ideal shapes, sizes and skin types so that they have easily identifiable yardsticks against which to judge their products. But I don't believe the judgements they pass are valuable, or even valid for modern women.

We need a redefinition of beauty from the beauty business – an endorsement of character and diversity rather than the promotion of a physical ideal. I think it is much more appropriate to take a holistic view, to look at the whole, body *and* soul, spirit *and* character.

DON'T STOP!

The beauty industry has done us all a terrible disservice by emphasizing just two elements: appearance and chronology. But we all know how easy it is for youth and looks to become tyrants. Women so often feel they've got to make an impression with how they look because that is how they will be judged. But I meet so many women my age who are throwing out that line of thinking. They're saying: 'Define me by things other than by what I look like – consider my wisdom, my humour, my grasp of what really matters in life.'

I think one big problem with the concept of maturity as we've been handed it is that it presupposes a fixed state of being. It becomes an excuse to stop. In fact, we never stop being given opportunities to change and grow, and that's a fact the ageing world is waking up to. As usual, statistics tell their own tale. In the UK, women aged 45–55 now go to the movies 55 per cent more than five years ago. Health clubs have had a 55 per cent increase in the number of 45–49 year olds over the past 10 years (compared to 25 per cent for 25–29 year olds).

All of this underpins the fact that the best kind of beauty implies self-respect, joy, wonder and inner strength. It isn't what you see. You can sit with the most beautiful women in the world and be bored out of your head in five minutes. In a vivacious woman, not necessarily a pretty one, her personality, charm and character can shine through. I believe true beauty has more to do with inner harmony than an idealized arrangement of physical features.

I'm 62 and I am what 62 looks like. My hair is thinner, the shape of my body has changed and I have lines on my face. I've found growing older has been quite disarming, which has been a surprise given that I've always been so scared of the final curtain. But I've also always loved changes and ageing has been like a sequence of little beginnings. The energy and curiosity that serve you well when you're a kid don't lose their power as you get older. Sometimes I think I've actually got more in common with my grandchildren than a 'mature' person my own age. So I won't be pressured into becoming obsessed with the ageing process: I intend to carry on loving every step. Women of my age still want to be seen as sexual beings, but we are not blindly attached to youth. Like most women in their fifties, I am more concerned with being heard than what I look like. I am not here to persuade visually, I am here to persuade by what I say.

WHAT PEOPLE REALLY THINK

The stance of The Body Shop on beauty was endorsed by a Gallup poll in Britain. Over half (53 per cent) of the women surveyed chose 'a sense of humour' as the most attractive characteristic of a woman. This was followed by 'enthusiasm' (15 per cent) and 'intelligence' (12 per cent). Self-respect, a sense of humour, wisdom and intelligence all rated substantially above good looks and a good figure. These results suggest why we see so many more women – young and old – prepared to confront stereotypes and become more active at the grassroots level, in local politics maybe, or campaigning on behalf of the environment, abused women, the homeless or any other of the key social issues.

Apart from looking at women's attitudes to themselves, the survey also looked at their attitudes towards the cosmetics industry. The results made it obvious that the beauty business just doesn't understand women. **Women clearly believed honesty is the best policy for cosmetics companies, not only about the ingredients in products but also about realistic product claims.** This concern with honesty was underlined by the fact that nearly 70 per cent of respondents believed that the use of models in cosmetics advertising can mislead people about the benefits of products. A similar majority felt that women are inaccurately represented in advertising.

A GLOWING FUTURE?

I have no doubt that science will one day stop the ageing process. Arthur C. Clarke reckons that we'll have unlocked the secret of immortality by the end of this century. The author of *2001: A Space Odyssey* has a track record on predictions that is pretty good, so I'd be prepared to take his word for it even if I hadn't read an eye-opening article in *Wired* magazine about the current state of research into life extension. But what I will *never* believe is that the antidote to ageing – whether you live to 100 or 1,000 years old – can be found in the 'miracle' ingredient of a dream cream.

Even when anti-ageing technology makes the length of time you stick around just one more life choice – like whether to quit your job or leave your partner or have a kid – I think people will still be asking themselves this question: what is the best way to age? Do I fight it or embrace it? And when I look around me these days, I see women in their forties and fifties coming up with much more interesting answers than a facelift.

Looking back on the twentieth century, schisms seem to define the age – including schisms between mind, body and soul. Perhaps that is why the new century rides in on a desire to put things back together, re-establish old connections, rebuild communities. I hear a lot of talk about an emerging integral culture, one which connects people and nature, flesh and spirit, and communities with other communities all round the world. I'm all for it. Perhaps then we will be more interested in what goes on inside the mind than outside the body, and more interested in responses to ageing that are creative rather than cosmetic.

So let's spend the money on a great bottle of wine instead and enjoy a great conversation with friends – and laughter. The laughter of women together is so revealing. It is the recognition of freedom and friendship.

And as we age, we will get more radical: that's the pattern. Remember what Dorothy Sayers said: **'An advanced old woman is uncontrollable by any earthly form!'**

BECAUSE WOMEN'S WORK IS NEVER DONE

AND IS UNDERPAID OR UNPAID OR BORING OR REPETITIOUS AND WE'RE THE FIRST TO GET THE SACK AND WHAT WE LOOK LIKE IS MORE IMPORTANT THAN WHAT WE DO AND IF WE GET RAPED

IT'S OUR FAULT

AND IF WE GET BASHED WE MUST HAVE PROVOKED IT AND IF WE RAISE OUR VOICES WE'RE **NAGGING BITCHES** AND IF WE ENJOY SEX WE'RE **NYMPHOS** AND IF WE DON'T WE'RE **FRIGID** AND IF WE LOVE WOMEN IT'S BECAUSE WE CAN'T GET A 'REAL' MAN AND IF WE ASK OUR DOCTORS TOO MANY QUESTIONS WE'RE **NEUROTIC** OR PUSHY AND IF WE EXPECT COMMUNITY CARE FOR OUR CHILDREN WE'RE **AGGRESSIVE** AND 'UNFEMININE' AND IF WE DON'T WE'RE TYPICAL WEAK FEMALES AND IF WE WANT TO GET MARRIED WE'RE OUT TO TRAP A MAN AND IF WE DON'T WE'RE **UNNATURAL** AND BECAUSE WE STILL CAN'T GET A SAFE, ADEQUATE CONTRACEPTIVE BUT MEN CAN WALK ON THE MOON AND IF WE CAN'T COPE OR DON'T WANT A PREGNANCY WE'RE MADE TO FEEL **GUILTY** ABOUT ABORTION AND ...

FOR LOTS AND LOTS OF OTHER REASONS WE ARE PART OF

THE WOMEN'S LIBERATION MOVEMENT...

WOMAN POWER

MAKING IT AS A WOMAN

At the end of 1999, I received an extraordinary letter from a member of staff at The Body Shop at Home, the new direct sales organization. She told me a heart-rending story about how her life had unravelled after her daughter had been killed by her partner. The man was out of prison just three years later and came to live just a few miles away from her. Not surprisingly, the whole traumatic experience gave her panic attacks, completely destroyed her confidence and self-esteem, and eventually she found herself losing the will to live. The letter meant an enormous amount to me, not just because joining The Body Shop at Home had helped her, but because it reminded me how loss of self-esteem can paralyse everything that once seemed worthwhile in life.

Most of us, thank goodness, don't have our lives torn apart in such a brutal way. But loss of self-esteem is such a common modern problem and it particularly affects women. In its extreme form it can kill and even in its milder aspects it can still strip away joy and make people powerless to change their lives.

'I will never know whether it was sheer madness or whether some greater power guided me,' the woman told me, describing how she had recovered from a suicide attempt and found herself accidentally at a Body Shop party:

Poster on the wall of area at The Body Shop headquarters in Littlehampton, under the headline 'Because We're Women'

Within a month I had signed the forms to become a consultant and paid for my kit. Suddenly I was terrified. What had I done? I couldn't do this job, I didn't have the confidence! What about the panic attacks? However, I did it. Within six months I had noticed that I was becoming myself again – my <u>old</u> self. I was surprised one evening at a party when I realized that I was laughing, really laughing. I had forgotten what it felt like. Within a year, I had my confidence back, my self-esteem, my sense of humour. I had my inner strength back and my old ambitious nature had resurfaced.

Stories like that are one of the things that make work worthwhile. But the point was not just that work combined with fun can rebuild people's sense of themselves, but also that self-esteem is a precious life-enhancing commodity.

The history of the past few centuries has done little to repair the self-esteem of women and in our own generation there are so many pressures to undermine it – a beauty industry making women dissatisfied with their bodies, an economic system that is often stacked against them, a set of hierarchical traditions designed to exclude them. But paradoxically, it looks as though the next few decades will need women – and so-called 'feminine' strengths – more than ever before.

I treasure the company of women. I love their laughter. I am astounded by their ability to keep communities together around the world.

My adoration and awe for the power of women has a rich personal history and stems from having a formidable mother who taught me to challenge everything. Such wisdom, passed to a child by a parent, bears comparison with anything in the Bible. From day one I was taught that I had a bold spirit and could create a world that allowed that spirit to flourish.

Many women, however, are not taught these values and if they are at all bold or enthusiastic, they are seen as shrews or eccentric. Naomi Wolf, an impressive young woman, beautiful, bright and intelligent, was slammed by the press because she was attractive and yet was challenging the cosmetics industry. Apparently you can't win – if you look like a dog you are criticized and if you are attractive you are criticized too. And the saddest thing is, the criticism often comes from other women.

That wonderful journalist Bel Mooney suffered at the hands of her sisters in journalism in the same way. Several years ago she said: 'All I am trying to do is stop a bloody motorway going through Bath and do I get support? No.' She went on to say that the people who gave her least support and who challenged and ridiculed her the most were other women journalists. I really felt for her. It is almost as though women feel they have to display a lot of bullying male characteristics to prove their worth.

The clichés claim that women bosses run a caring, sharing shop, but, as already mentioned, according to a Manchester Business School study the truth is otherwise. The survey says women at the top are tough cookies, autocrats who rule by fear. That may be a misreading of the entrepreneurial spirit, regardless of gender, yet it may also be that women are overcompensating.

In spite of their best efforts, I don't believe there is much chance that the proverbial 'glass ceiling' is going to be shattered in my lifetime. Some reports indicate that if women continue to progress in business at the current rate, it will be 500 years before they have equal managerial status in the world, then another 475 years before they hold equal political and economic status to men. Nearly 1,000 years to go isn't progress. But I also believe that there are ways to work around the problem and my own career shows it's possible – though even then the traditional male structure is still in place waiting to reimpose the status quo.

Unfortunately, around 40 years after the women's movement started – one of the most unheralded movements in history – women's magazines are still full of advertisements bombarding us with images of smiling and compliant women whose message seems to be: 'I know my place and it is here giving female support to a male breadwinner in the kitchen, submissive and unthreatening.'

And the emergence of the so-called 'new man', who has hijacked the role of caring, sharing male and is lauded for his sensitivity to his partner's needs and his responsibilities as a father, just makes me laugh. How is it that the new man manages to garner more romance and glamour in that function than women have ever done?

WHY THE WORLD NEEDS WOMEN

IT'S THE ECONOMY, STUPID

Legalized job discrimination against women supposedly ended in Britain in 1976. That did much to remove obvious discrimination, but it didn't deal with the ideology of pin money – the assumption that women aren't working for real, that their money doesn't pay for basics like bills and food, only disposable extras like holidays or hairdos. From that biased assumption there grew a whole structure of lost opportunities for women in our society.

Yet the work that women do has always been fundamental to the global economy. For thousands of years, all over the world, woman have been traders, farmers and entrepreneurs – indeed, whatever they have had to be. But their contribution hasn't registered with traditional economic institutions because so much of it has been non-monetary. The women economists in the team that put together the new methods of economic measurement after the Second World War weren't even listed in the acknowledgements. Nowadays, one common economic term for non-monetary work is 'inactivity'. Our national accounts assume that because no money is changing hands that nothing is happening. It's that attitude that has made women's work invisible. No wonder the battle cry of the women's movement was 'Equality!'

We need to press for greater gender equality on a massive scale.

When most people talk about human rights, they mean political and civil rights, things like freedom of speech, freedom of worship, freedom to own property. I agree these rights are essential; indeed, at

The Body Shop they are frequently promoted in our campaigns. But I have to say these rights tend to be masculine rights. **Read the Universal Declaration of Human Rights and you'll find other rights which aren't referred to nearly so often. They include the right to a family, the right to rest and leisure, the right to an adequate standard of living, the right to a cultural community.** These basic economic, social and cultural rights address the particular concerns of women. They are as fundamental a right as free speech. Just how fundamental to society is obvious from the fact that it is women who universally hold a community together, who fetch the wood and water, who cook the food and care for the kids.

And women's lives aren't getting any easier. Since the creation of the Universal Declaration in 1944, poverty has spread, social exclusion has increased and the chasm between rich and poor widened. Political and religious leaders seem to have arisen everywhere who view contraception and feminism as interlinked evils. And as global poverty expands with unfettered free trade, it hits women and children hardest.

Why does business have to work this way? Why not harness the market to eliminate poverty? Why not improve life for the world's poorest first? Is it so impossible to move business from private greed to public good?

We have the resources. I sense that in the growing vigilante consumer movement, we have the popular will and – God knows – there is plenty of inspiration in the small-scale grassroots initiatives that women have been so instrumental in establishing in the majority world.

The Babassu Example

Let me give you one example. Several generations ago, the Brazilian government resettled poor people in the north-east of the country, where they made a subsistence living harvesting babassu nuts from the floor of the forest. Then the cattle barons began moving in, fencing off enormous tracts of land and driving the settlers off by force. Hundreds of men died, but the women were smart – and brave. They knew the code of machismo prohibited the cattle barons from harming them or their children, so they formed their own non-violent resistance movement: Free Babassu. They held sit-ins round babassu trees and after decades of campaigning, eventually won the legal right to go back into the forest and harvest the babassu nuts again.

The Body Shop used to get the babassu oil we use in our cosmetics from a middle man, but when we heard about these women we wanted to see if we could deal with them directly and pass on the savings to them. They weren't set up to press the oil, but in 1994, with the help of a Brazilian NGO, we began sourcing our oil directly from a women's co-operative in north-eastern Brazil. The benefits were immediately visible. For the first time, local commissaries were well stocked with staples and foodstuffs, but more importantly, this trading relationship raised the profile of the women in the community, both as breadwinners and as a political force.

IN COMMUNITIES, TOO

Given the tools, women can rebuild communities. That's why I am keen on saying again and again that the strongest form of women's leadership is at grassroots level. A sense of community is one of the so-called 'feminine values' that ethical business thinkers put forward in their quest for new paradigms. These values reflect intimate personal and cultural attributes which are in many ways the reverse of the global market syndrome, which is all about distance, impersonality and the movement of capital regardless of human consequence.

We must urge governments and businesses to help women in need by supporting small-scale grassroots initiatives. We have to

Why shouldn't women make good coaches? We were brought up to listen, to nurture, to observe.

Billie Jean King

put them first, as leaders and advisers and active participants. We have to listen to their experience. Globalization is a mug's game being played in a man's world. I can imagine a day will come when compassion counts as much as cashflow. After all, the challenges that confront the business world already demand a holistic perspective – and women will certainly be best equipped to face that kind of future.

As a woman and a worker I share two universal experiences of women around the world – and two simple truths:

☛ It is still women who carry a double burden. Women in most societies, including our own, shoulder the responsibility for caring for the family, worrying about the household and being the one who drops everything else when a child is ill. In the Third World, they also usually grow the household food, fetch the wood and manage the water supply.
☛ In practically all societies, women are victims of either discrimination or disadvantage.

Mix those two truths together and you produce the situation already described, where the work women do is considered unimportant. So policy-makers, agricultural banks and aid agencies ignore women's work and deny them the fundamental economic activity required to feed themselves and their children. In doing so, they ignore the people who *really* count. I think we need to recognize the value of the work women do, especially their work in the home. We need new barometers to measure the full value of women's contribution to national economies, both in the developed nations and in the Third World.

Perhaps we should take a lesson from the Gabra, a nomadic pastoral-farming group in East Africa. Among the Gabra, when a man reaches a certain age, he crosses the gender barrier to adopt the rituals and roles of an elderly woman in his culture. He listens, tells stories, and weaves and endorses the emotional and spiritual fabric of community, the traditional function of women. This is the highest honour he can achieve.

Travelling in the developing world, I have seen how the poor rarely have access to the information and ideas that can help them

Work is and always has been my salvation and I thank the Lord for it.
Louisa May Alcott

escape poverty. They are also more likely to be female, particularly in urban areas, where they are paid less and work more – women receive only one tenth of world income but are responsible for two thirds of all working hours. They are also expected to give birth to, raise and feed numerous offspring, which saps their strength and puts their health at risk. Commonly abused and beaten, they have fewer legal rights and property rights than men. Even when we carried out a survey on domestic violence among our own staff at The Body Shop, we discovered that 16 per cent of our workforce was continually battered and 60 per cent knew someone who was battered.

But, despite all this, it is invariably women who keep communities together and I believe that today there is a return to community. It may not be the kind our grandparents knew, but it still means sharing common practices and acting as if everything we do matters.

AT WORK, TOO

By moving into the world of paid work, in rich countries at least, women have certainly raised their visibility. In some industries, a whole new class of female managers is emerging and transforming the workplace. Women managers are effective social initiators and their biggest strength is communication. Women build alliances, bring people together and, most importantly, develop intimate informal networks. These networks circumvent hierarchy through unexpected points of contact. They promote lateral thinking and the sharing of information.

However, I doubt that you could make a very conclusive case that women in the workplace have become equal to men. The media love female high-fliers, the handful of company directors and CEOs who are trotted out time and time again as evidence of gains women have made. But they are not truly representative of the average working woman, saddled with a double burden as she tries to balance her job with life as a mother and a homemaker.

It is almost as if women are conditioned to believe that they will never have a role to play in business. We women in business are actually measured by how many male characteristics we have. All

GOD
COULDN'T BE EVERYWHERE SO SHE CREATED MOTHERS

Women who were once thought to be inferior leaders because they were 'too emotional' now turn out to be excellent leaders because they can exhibit 'special' emotional qualities.

Harvard Business Review

this reflects the stereotypical male thinking that our emotions and our caring and our sensitivity and our intuition are not to be brought to the workplace. And yet it is these very ingredients that will change the notion of business.

The best reason for believing that more women will be in charge before long is a ferociously competitive economy where no company can afford to waste valuable brain power simply because it's wearing a bra. The problem is that most women don't know how remarkable they are and they don't have the self-confidence. Women of my age come from a culture that tells us to suppress ourselves, to glory in what our husbands have achieved, so any woman who is entering business has to reinvent herself. The way women do this is extraordinary, but it is never celebrated. Nevertheless, we've introduced flexitime, we've humanized the workplace, we've brought in daycare centres and we've brought in a notion of humanity, a notion of love. **We've made it legitimate to talk about love in the workplace.** And we've brought in the notion of creativity. Yet it is still not remarked upon because – to men – business is about finance.

I'm not talking about rocket science here; I'm merely talking about changing the language of business, so that humility, love, creativity, compassion and understanding become part of the lexicon of business dialogue.

There is a myth that women can succeed only by adopting male behavioural patterns. But it is only a myth. Research in the USA shows a different profile emerging, of women managers who are effective social initiators, who anticipate problems and present possible solutions. American women build alliances, bring people together and most importantly, as already mentioned, they develop networks. Their biggest strength is communication.

We have to rethink the whole management education and bring gender issues out of the shade and into the sunshine. Everywhere I travel, I encounter people, mostly women, who instinctively know how to manage this planet but are silenced by a very soulless world view that gives privileges to the white, the male and the educated. Truly engaged global management education must learn to enter into a loving and respectful dialogue with all voices.

Alex Elliot

future prime minister, inventor,
entrepreneur, philanthropist.

(shown here with her Brother Chris)

love **your**
body

AND AS ENTREPRENEURS

I also believe we need more women to go into self-employment and set up small businesses. These female entrepreneurs desperately need the backing to grow businesses and create the jobs they need for others. In every country I have travelled to in the West, it is the older, larger corporations that are dying of boredom and losing millions of jobs. Corporations as we know them were created by men for men, often influenced by the military model, on complicated and hierarchical lines and are both dominated by authoritarian principles and resistant to change. By setting up their own businesses women can challenge these male-dominated corporate models.

When I look at the businesses run by women today it is noticeable that there are no instruction manuals for them, because no one has ever done what they are trying to do before. All that gives us enormous freedom to experiment.

And behind all this experiment lies the real task before us; not just to give women equality on men's terms, but to create a new partnership of men and women designed around the best values of each of the sexes.

AND AS INTELLECTUAL LEADERS

In this respect, there is already plenty of evidence of the influence of women in the work of pioneering female thinkers whose concern about the society their children will inherit promises to fundamentally change global economics. I am thinking of women like Alice Tepper Marlin, the author of *Shopping for a Better World*, who was the first person to research, in depth, the social and environmental impact of corporations. Or of Amy Domini, lawyer and the author of *Challenges of Wealth*, the bible of socially responsible investing. Or Vandana Shiva, a physicist and philosopher as well as India's greatest environmentalist, who has led the environmental and social critique of economic globalization from a Southern perspective. Or, for that matter, futurist economist Hazel Henderson, whose books about the way things could – and should – be have crystallized the issues for ethical business people of both genders. **In fact, most of the financial sector's innovative thinking on socially responsible investing has come from women. Why am I not surprised?** And it's women, far more than men, who form the spearhead of the global stand to halt poverty and the destruction of our environment by founding small-scale grassroots initiatives around the world. I've had firsthand experience of some extraordinary ones, like the Chipko movement in the Indian state of Uttar Pradesh, which began with women responding to forest destruction by physically protecting trees and which is now that country's largest grassroots environmental organization.

It is interesting that women are at the forefront of environmental activism. I believe one of the changes we will inevitably see is a growing appreciation of what women have to give.

In their role as providers, women have a gut appreciation of the realities o human existence.

MOTHERHOOD

When my daughter Justine had her first baby, she thought I would be pressuring her to go back to work as soon as she could. She was wrong. I told her the most extraordinary time she'd have with her child would be those first few years and so she should take them. After all, I did – I had Gordon who worked from home and the support of the extended family that is a fact of life for an Italian. I also thought it was a good idea that when the children lifted their heads they would see and feel love from people other than their parents, so this created no problem for me when I started The Body Shop. At that time my mum was there to help me with the children – Gordon had decided to go off for a couple of years to ride a horse from Buenos Aires to New York in his pursuit of the grand gesture.

There is a theory that working mothers breed behavioural problems in their kids, that it is quantity time, not quality time, that counts. But I don't know one woman who doesn't need to work. So the working mother is damned if she does, damned if she doesn't – a familiar and frustrating situation. Blaming mum is usually a sign of refusing to grow up in a man, so I suppose it makes perfect sense that our male-dominated society still sees fit to blame the mother, whether she is a single parent or with a partner. It's still true that men spend seven times less time with their kids than women, so equality in the home and the workplace is just a myth.

Because of this, Britain continues to have a gender-segregated job market – in very general terms women are doing the low-paid domestic-type people-interfacing jobs while, on average, earning about three-quarters of the hourly wages of men. Many women need to work part-time because it is the only way to cope with their family needs and responsibilities. And at the moment, part-time jobs are, by their very nature, usually lower skilled, poorly paid and unlikely to provide opportunities for promotion.

I think there are three major solutions to this problem and I believe we must pursue all of them more meaningfully than we have done thus far. They are:

- childcare provision in the workplace
- upgrading part-time work
- redesigning the working world to become family-friendly

I believe we need to draw up a blueprint that accepts motherhood as an integral part of our working lives. In my rosy vision of the future, business will acknowledge its responsibility towards protecting the family. So when governments fail to support working parents by not providing daycare facilities, business will create a special place where the parents are served and child development is supported, where families are welcomed and values are explored and protected.

Providing good-quality childcare is already an absolute priority so that working mothers and fathers can see their children gain a valuable education and social experience while they earn their living. Childcare has been a low-status activity in this country, but – as some economists say – all our futures depend on the next generation. That means anyone who takes no part in bringing them up or educating them is getting a free ride and we are morally impelled to recognize childcare as one of the most important tasks there is.

As an employer, The Body Shop took an innovative first step towards resolving this issue in 1990, when we opened our workplace Child Development Centre in Littlehampton. It cost about £1 million, and has places for children between the ages of three months and five years. It also runs an after-school scheme, where children are collected from local schools by nursery staff and are involved in various activities until their parent finishes work, and offers day camps for children from five to eleven during the school holidays. We also offer places to other local employers and emergency places for social services and we operate a voucher scheme to subsidize the childcare arrangements for our staff away from Littlehampton.

Our Child Development Centre is the centrepiece of our family care policy and the symbol of the direction in which we wish to travel as an employer. It is a message to our staff that we think their children are important and that we don't downgrade our commitment to our staff – men or women – when they become parents. I also think the centre is the most fascinating part of our complex to visit because there is such an authentic high level of energy, love and life.

And for those who think all this education is too expensive, remember one of the slogans that appears on our lorries: 'If you think education is expensive, try ignorance.'

THE POLITICS OF RELIGION

While we're looking at the subliminal ways in which women are controlled, we also need to look at organized religion. Religion today is less about the meaning of worship and seeing the splendours of the divine in all living things than about control.

Anytime we see God as the male, rather than seeing God in all of us, we are cementing traditional thinking.

We badly need to look at the control politics of religion and subvert it. When we put up a window poster saying 'God couldn't be everywhere so she created mothers', it practically caused a riot in the shopping malls of New Hampshire. New Hampshire is, after all, an American state where the word 'imagination' has been banned in many schools which are controlled by the religious conservatives. Added to this, you have the right-wing pro-life lobby determined to seize control of women's reproduction as well.

FINDING OUT ABOUT AMERICAN WOMEN

In America, The Body Shop was part of an enormous groundbreaking initiative to survey working women across the 50 states. Women's

There is a saying: 'Where power is, women are not.' Women must be willing to be powerful. Because we bear scars from the ways men have used their power over us, women often want no part of power. Petra Kelly groups, companies and the US Department of Labor asked women about their working lives and what would make their jobs and their workplace better for them. Through their 200 stores across the US, The Body Shop helped distribute the survey questionnaire 'Working Women Count'. The response and enthusiasm it generated were extraordinary. The results constitute a unique resource bank of women's concerns, problems and ideas on employment issues.

It immediately became clear that the survey had struck a deep chord with women, who seized the chance to describe the reality of their working lives. **In the space on the questionnaire form reserved for a message for President Clinton, one woman simply wrote: 'I'm tired.'** This echoed the answers of a quarter of a sample taken out of the initial responses. American women wanted to let the American government know they were fed up with the exhausting task of juggling work and family duties with little help.

I'm sure that millions of British people would echo these sentiments. This also includes Britain's working men, who work the longest average hours in the European Union, many of whom would dearly love to have shorter hours and opportunities for paternity leave. What a shocking contrast this all shows up: we see an American government that opens itself up to comment and criticism from working women and a British government that falls short of a meaningful commitment to something as basic as a universal nursery provision.

EMPOWERMENT

So The Body Shop will carry on supporting women's organizations in their calls for better employment law and practice for proper childcare, and equality and fairness in every aspect of life, including politics. Our fundamental approach has been one of empowerment. Women in Britain have a long and powerful history as campaigners for justice and human rights. We are giving them as much support as we can and helping women's organizations reach a wider audience. That's why in 1994 we helped publish a remarkable book called *What Women Want.* This book brings together the policy agendas of over 70 key women's organizations and women activists in Britain and their simple, rational and clear demands for change.

We will continue to take our place in whatever discussion is essential for action and we will demand the right to have everyone's contribution matter, whether it is in marriage, politics, friendship or work.

My own experiences as a businesswoman have shaped my thinking on this. While I am simply not able to communicate in the arcane and abstract language of social policy, economic initiatives, per capita incomes, demographic data and all the rest of the jargon, I can talk with feeling and authority about the concrete realities of the role of women today – in life and work, in the developed as well as the undeveloped world.

As I wrote in an earlier chapter, the community trade programme of The Body Shop plays an important role in setting up small-scale economic initiatives in economically deprived communities around the world, exemplified perhaps by the women's co-operative in Ghana which provides The Body Shop with shea nut butter. Dealing directly with economically marginal communities and co-operatives around the globe, I've seen how women hold a society together. Economic opportunity means much more to them than money. It promotes fundamental self-esteem, facilitates education, healthcare, cultural continuity and the chance to protect the past while shaping a future. **Women simply have to be listened to.** The fourth United Nations Conference on Women, held in Beijing in 1995, was marred by shambolic organization and lousy weather, but it was still a call to the world to listen to women. Since the previous conference, 10 years earlier, in Nairobi, little had been done. The conference call then was equality

No Shit!
Dear, It says right here that men and women are created equal.
EDGAR'S BIG REALIZATION.
THE BODY SHOP
'A Registered Attitude Company.'

by the year 2000. It is clear that these years have been largely wasted for women world-wide.

Meeting our demands for change is the task of politicians and activists. The beneficiaries will not simply be the women of Britain, on whom the last few years have been exceptionally hard. Is it possible to imagine Britain as the greatest champion of women's rights in the world? I yearn for this country to be remarkable at something, to be at the head of the pack, leading the crowd. We should not be a second-tier nation in terms of women's rights.

SELF-ESTEEM IS THE KEY

Self-esteem is a much despised and dreaded word, often reinter-preted as selfish and self-indulgent by those who depend on women's self-sacrifice. But it is also the route to revolution. We are not in the habit of making the connection between self-esteem and democracy, dignity, political activism and freedom of sexual expression, but in the future we will be. In short, self-esteem is the root of finding our way in the new millennium.

Self-esteem starts in childhood. It starts with the young girls who need our support, who need to rely on the wisdom born of experi-ence we can give them, not what the media tries to feed them. By not teaching young women or young men to develop their own voice and their self-esteem as they enter the years of puberty, by omitting gender issues from the curriculum, we are dealing them a huge injustice. I believe schools should teach our children about relationships, respect and non-violent behaviour. We should be encouraging our children to take back the images of their bodies, to know their legal rights and how to identify the myths and stereotypes surrounding human behaviour.

Then there's the 'invisible' education – the education given to young girls by organizations like the Girl Guides and YWCA. This education is often overlooked, yet the impact of it can be terrific. When young girls can assert themselves confidently in the world, they can also apply and assert their own values: the values of kindness, co-operation and sharing — the values we need to reform and reshape our world.

We can't have self-government without the self-confidence that is at the root of it.

And what of that other invisible education: the education in the home, in the family? We know that the behaviour children learn from their parents affects them for life. We know that children who are abused often continue to treat themselves badly, and we know the root cause for both girls and boys is that same lack of self-esteem that has enabled the media to alienate people from their own bodies. The beauty business has a lot to answer for. Changing it couldn't be more urgent.

Get Out of the U.S!

TAKING A BATH IN THE US

I should have realized, when a shopper in New Hampshire fainted at the sight of one of our bawdier posters, that the expansion of The Body Shop in the US would have its teething problems. The poster showed a bronzed muscular man with a bottle of Watermelon self-tanning lotion – called 'Fake It'! – stuck in his pants like a giant willy. It was blunt, certainly, but it was fun and it gave people a laugh when our art department developed it in London. In Australia, it was so popular that they put it on T-shirts, so when it produced tidal waves of hysteria in the US I have to admit we were a bit surprised.

I had to defend the poster on TV against the accusation that it was sexist. 'At a time when Americans are debating double standards in sexual stereotypes, here's another example,' I said. 'Everywhere you turn, ads are using sexual titillation to sell products. Put up a man and people are offended. That's silly. This is just a light-hearted pun on our modern obsessions with beauty icons – but since it is about virility, it arouses more interest.' What I didn't say was that while the protest was going on, we were ordering an extra print run to meet customers' requests for a copy of the poster.

The story even made the front page of *USA Today* under the headline 'Mae West would have been amused. Others are not.'

Poison pen letter we received after our Fake It! campaign in the US

FAKE IT
Self-tan lotion

THE BODY SHOP

I thought it was hilarious. Apparently you can have breasts tripping out of cantilevered brassieres, but not this. But then that's America. It's different there.

Viewed from Europe, the United States is the graveyard of European retailers. One European retailer after another has crossed the Atlantic full of hope only to flounder in the waste-land of shopping malls. The Body Shop was very nearly interred in America too. We made a whole slew of major mistakes, partly due to our arrogance and partly due to huge *naïveté*. We did not anticipate the problems because we didn't think ahead. We were still thinking like entrepreneurs, pushing an idea and seeing how far it would go. We thought of ourselves as a shop-opening company. We'd opened up in America, we'd opened up in Japan – how cheeky we were! We were sort of like Alice in Wonderland, growing bigger and bigger – but we never looked over our shoulders to orientate ourselves. No analyst could understand how we could open 160 new shops around the world every year, but for us it was as simple as breathing. Or at least we thought it was.

When we entered the US market in 1988 we opted to open our own stores rather than immediately get into franchising, to give us time to adjust to the new market. Retailing space in the US is controlled almost exclusively by shopping malls, where most of the trading is done. This is a development which I believe has done more to destroy Main Street America – and consequently community spirit – than anything else. They say they will give you one good space in a mall to open a shop if you will take three crummy spaces in other malls. We didn't want to play that game, but we soon found ourselves sucked into it.

I find the ideas of keeping people in one area for four or five hours very unappealing – and the malls are controlled by big retailers rather than small shopkeepers. Still, we started in what I think was an intelligent way, casting around New York first just to see if we could make money, which we could. But then we were seduced into believing what the press was saying about us, seduced into thinking that this was the way it was always going to be. Our problem was that, in a range of different ways, American retailing was different.

America is not a melting pot. It is a sizzling cauldron.

Barbara Mikulski

THERE WAS THE COMPETITION...

We thought when we entered America that we were sacrosanct, untouchable and that we would be able to open up an entirely new market with barely a second thought. How wrong we were. What we didn't anticipate was how incredibly fast the competition would come in.

In autumn 1990 – the same year we started franchising – retail mogul Leslie Wexner opened the first Bath & Body Works, the shop that would soon be described as our main competitor. Within 18 months, he had 100 stores grossing $45 million a year. Within another 18 months there were around 30 different lookalikes of The Body Shop and we were struggling. Competitors like H_2O Plus, Goodbodies, Origins and Garden Botanika jumped on the natural products bandwagon, developed their own lines of fruity potions and sold them for less. It was then that we began to understand the grave implications of the copycat marketing methodology in America.

But then, we *were* a shop-opening company. What we didn't have was a strategy to cope with competition that was so fast and furious. We had not thought of what might happen if a big retailer like Wexner, who rents an enormous amount of retail space in the malls in America, suddenly decided to turn one unit in each mall into a copycat of The Body Shop. Immediately, we had 500 shops competing with us. We never even thought like that.

We imagined people could separate the authentic from the imitation, the real from the frivolous. But in America that kind of discernment was absolutely irrelevant.

AND THE MARKETING...

The fact that we were opening so many shops masked the reality of what was happening. While overall sales were rising, many of the shops were struggling with static or even falling sales.

In the rush to expand, we didn't pay sufficient attention to marketing details. Small irritations became significant. The name of the men's range, 'Mostly Men', was unpopular; we eventually dis-

Thanks
for
the
dream
Dr King

covered 'The Body Shop for Men' was preferred. Similarly, Americans like their shampoos in flip-top bottles, not screw tops. The fact was we hadn't understood the language of American retailing. **The trouble was that our whole ethos made us virtually unmarketable.** Concern for the environment, a refusal to test ingredients on animals and as a result a determination to pursue community trade proved to be enormous handicaps in the USA. There were so many things we couldn't do, from manufacturing to distributing. We couldn't use about 50 per cent of the available ingredients because they had been tested on animals. We were bending over backwards to find products capable of being sourced in the Third World and harvested by indigenous communities, so we were much less able to respond to the demands of the marketplace. America is the ultimate consumer and it kept asking for new products which we simply didn't have the means to supply.

AND THE MARGINS...

We also didn't appreciate the implications of the huge margins in toiletries in the United States. Franchising ate up those margins. The way our business was structured, head franchisees ran networks of sub-franchisees. The head franchisee needs to have a margin to organize training and getting the shops operating, and the sub-franchisees have to earn a living too. This extra layer meant that our margins were drastically reduced, while our competitors were buying products wholesale for 10 cents which they could retail for a dollar then offer a 20 per cent discount and still make a big profit. American consumers expect discounting and we couldn't give it to them. **We also didn't understand the concept of never-ending sales – we had never ever held a sale!** And we had never come across their prevailing strategy called GWP, gift with purchase. American consumers were also accustomed to the idea of being offered two for the price of one, something we had never done either. The upshot was that we found ourselves trading in a discounting culture that was entirely alien to us.

And we soon discovered that everything we thought was quirky and different and clever about The Body Shop was much less appreciated in the United States. By then we were quite accustomed to introducing politics into retailing and so we didn't think twice about using some of Martin Luther King's speeches to commemorate Martin Luther King Day in our shops. It proved to be tremendously unpopular with the mall owners. Mall owners in America don't like confusing politics with shopping. They gave us hell again during our voter-registration campaign and during every one of our human rights campaigns.

AND THE HUMOUR...

Before the Fake It hysteria, we produced a Christmas poster showing Santa Claus in a back-to-front baseball cap and trainers listening to

a Walkman and were accused of inciting 'gang violence'. Another poster promoting male deodorant was thought to 'encourage homosexual activity' and others were denounced – mainly by the religious right – as 'disgusting', 'immoral' and 'suggestive'. And to make matters worse, we made a genuine cultural mistake with a range of mother and baby creams which we called Mama Toto. We discovered too late that *mama toto* means 'motherfucker' in some parts of Mexico.

When we hung a banner outside our store in Castro Street, San Francisco, reading:

'2-4-6-8, USE A CONDOM OR MASTURBATE'

our then American CEO made it clear he was very unhappy about it. His antipathy made me feel dysfunctional in my own company. I would have been very disturbed if it wasn't for the fact that I realized a long time ago that dysfunction always occurs when the ideas of founder-entrepreneurs grow so big that they find it hard to identify with what they conceptualized.

AND THE ADVERTISING...

To add to our troubles, we were a company that did not advertise. In America, the spiritual home of advertising, that was considered to be completely mad and so we broke our own taboo and dipped a toe into advertising, but we were hopeless. We didn't have the courage or the history. Every time we tried it, it was more pathetic than the time before.

NOT TO MENTION THE PEOPLE...

When we decided to start franchising in 1990, we had more than 2,000 applicants. We gave them all questionnaires that asked them about their taste in music, books and films as well as their thoughts on things like how they would like to die. I thought standard job interviews were so boring and I wanted people who were politically aware and who were looking for a livelihood that was values-led, but we were accused of trying to form a cult of socially like-minded thinkers.

I think one of the quirkiest things we did was to say we weren't especially looking for business skills. It may have been charmingly innocent; it was certainly naïve. But I didn't want rigid thinking – I wanted people with passion and commitment. It seemed like the right thing to do at the time, but looking back it was a mistake. We got a lot of applications from activists, teachers and environmentalists who wanted to own a branch of The Body Shop because of what we stood for. Some of them turned out to be amazingly good, but some were so inept that you could write a sitcom about them. **I don't think there was anyone who came to us in that first year who was even remotely interested in the obsession with the bottom line – what they were looking for was an honourable livelihood. They wanted to change the way things were and to be part of a social experiment.** In a way this was great, but it meant we hired managers without any business skills whatsoever. Neither did we hire the right people as employees. We were completely seduced by the Americans' ability to be eloquent about themselves. They came across as so dynamic and charming, but too often it was all seduction and no delivery. We found we had often employed bullies or people who only wanted to work for us because there was a great golf course nearby. We also found we had top personnel orchestrating the worst possible decisions.

The Bullying Problem

In fact some business student should write a case study on our unerring ability to consistently employ the wrong people. We didn't know how to look for the signals of bullying, the signals of verbal violence, the signals of indifference to our reputation or mediocrity. We didn't even know how to detect indifference to retailing. And it was simply because we no longer practised due diligence in checking out behaviour, as we used to do initially. By then we were leaving that to consultants.

There was one particular director who worried me. The first thing he did to irritate me was to try and fit our marketing into a series of formulas. He introduced a notional customer whom he called Betty. Everything had to be pitched towards pleasing Betty. I got to hate Betty and everything about her. I find it difficult to like the name even now. I used to wonder, as I listened to this claptrap, where this thinking came from. I'm just delighted when any customer comes into our shops – Tom, Dick or Harry, young or old, anybody at all we can tell our stories to. I remember in the old days

This is ... a trait no other nation seems to possess in quite the same degree that we do — namely, a feeling of almost childish injury and resentment unless the world as a whole recognizes how innocent we are of anything but the most generous and harmless intentions.

Eleanor Roosevelt

sitting down with our staff and saying I wanted our products to be bought by anyone and everyone.

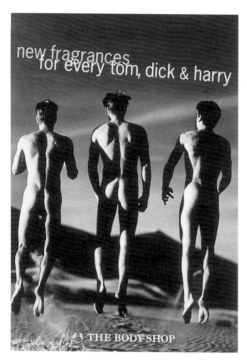

I'll never forget the time we were having dinner in San Francisco with two or three of our employees for whom I would lay down my life. They had been with me more or less since they were teenagers and I had mentored them for over a decade. I'd watch them grow up and they were just monumentally great people. They knew the touch and smell of The Body Shop better than anybody and were consistently challenging the director who foisted Betty onto us. To my horror, and within my earshot, he suddenly turned round to one of these people and said: 'If you don't toe the line I am going to fucking kick your ass out of here.'

I was absolutely furious. When we were alone I said to him: 'I don't like you, I don't like what you stand for and I don't like your bullying. I have no idea how you are at home, but if this is the way you are here, then I don't want it. I don't want any part of it – it makes me physically sick.' And do you know what? He thanked me for sharing it with him! What can you do? He lasted another two to three months after that.

AMERICAN HEROES

Fortunately the Betty director is not typical of America as a whole. In fact, what is? I am always entranced, seduced, mystified, infuriated and challenged by America. I love the enthusiasm there, the willingness to take risks, to experiment, and some of our most enterprising and amazing franchisees are American. I love the politics of dissent in America, its labour movement history and its community movement. I have also met more heroes in the United States than in any other place in the world.

I am thinking here of institutions like the Social Venture Network, magazines like the investigative *Mother Jones*, people like Carl Jensen, the founder of Project Censored, who for the past 29 years has worked to expose media manipulation by the US government, military and corporations. Every year, Project Censored publishes a book called *The News That Didn't Make News and Why*. Using impeccable research and unimpeachable sources, Carl was among the first to identify acid rain, the problems associated with decommissioning nuclear reactors and the US involvement in training death squads in Central America. But there are four Americans in particular I admire and am proud to be associated with:

Ralph Nader

Ralph is America's best-known consumer activist. He is the man I am most in awe of and more than anyone or anything else he has shaped my thinking on the politics of business. All his life he has played David to the corporate Goliath, proving how wrong multinational companies are when they use their size to silence the opposition. He has also proved how much difference one individual can make. He is the most feared man in corporate America and his obsession is truly frightening. As driven and reclusive as he is, he shows up in opinion polls as one of the few people in Washington that ordinary Americans trust and it's not hard to see why. He poses the simple question we all need to be asking: why shouldn't business be as responsible as an individual is expected to be? The fact that the corporate world is so amoral is an obscenity to Ralph and his undiminished capacity for outrage inspires me.

But I'm also inspired by Ralph's faith in the citizen as an agent for change at the heart and soul of the political process. People power works. We've seen it on the streets of countless countries in the closing years of the twentieth century. And in the corporate world, the all-consuming bottom line is also an Achilles heel highly vulnerable to people power.

Matthew Fox

Matthew is a theologian and visiting him is a high point of any trip I make to California. Defrocked by the Vatican, deemed dangerous by the Pope – whom he compared to the alcoholic father of a dysfunctional family – he is a director of the University of Creation Spirituality. Based in downtown Oakland, the Institute is remarkable for its pragmatic approach, embracing the body as it enhances the soul.

Matthew believes many of the self-esteem problems that plague society today have to do with our divorce from the world of ritual. It is in ritual, he says, that 'communities come together to heal, to celebrate, to bring out people's talents, to unite generations in a common origin story and therefore in a common morality'.

The kinship I felt with Matthew when I first met him in 1993 probably had an immediacy that sprang from our reputations as butchers of sacred cows in our chosen vocations. His speciality was spirit; mine – as founder of a cosmetics company – was clearly flesh. But we were good at making connections between things that seem contradictory and we both wanted to honour the soul of life. Wasn't that what my mother had tried to teach me?

I remember when Matthew came to talk to our head office in Littlehampton, his visit coincided with the scandalous goings-on of one so-called 'rave vicar' in Sheffield. Among the staff who jammed into our corporate cinema for the talk were a number of born-again Christians. I think they half expected to be greeted by a minion of Satan. Instead they got Matthew talking about the great mystics and how they saw God in everything. It was a lyrical, loving address and the audience was mute with admiration.

Matthew is one of the most influential theologians in the world today, but he is also very involved in the secular world. We once shared a stage together and talked about a subject very close to the heart of The Body Shop – the necessity of reinventing work by attaching a values system to it. Matthew believes that a sense of community has been missing from the worlds of business and science, but also that the so-called 'value-free' objectivity of these worlds has now been shattered. 'In science itself,' he said, 'there is a realization that participation is part of the way we interact with the world. And so business can no longer pretend to be objective, just measuring things by a few chosen quantitative criteria.' If he could advise us to pursue any one course of action, it is this: look to our hearts and souls. It is there that the real work of business has to be done today, because it is the responsibility of business to address the world in crisis.

Rebecca Hoffberger

Rebecca Hoffberger is the founder and president of the American Visionary Art Museum in Baltimore. Her passion is outsider art, the art made by the misfits on society's margins – poor, unhinged or just plain obsessed. There is a directness to the art shown at AVAM that is often quite breathtaking. At the same time, there is often a real grandeur, the grandeur of the human spirit in full, free flight.

It is through Rebecca that I have met artists like Leonard Knight, a Korean war veteran who painted a mountain in Niland, California. Actually, he practically made the mountain himself out of adobe. Now, baked by the sun, sealed with many coats of paint, Salvation Mountain can withstand anything nature throws its way. A primary-coloured riot of waterfalls, flowers and Biblical lore, it has inevitably become an attraction, with hundreds of visitors stopping by or even settling there permanently in their camper vans. Museums, art critics, sociologists and filmmakers have also come to see one of America's most extraordinary outcrops of visionary art.

Rebecca lives her life surrounded by people like Leonard. I could almost envy her for that, except that envy is hardly the appropriate response to someone whose earthiness and sense of humour seduce everyone she meets. Her charm was a valuable ally in her decade-long campaign to get AVAM off the ground. Equally irresistible was her incredible, almost steely drive. When she was 17, she told a reporter of her devotion to creativity and her concern for the people society overlooks. So AVAM is literally her life's work.

Rebecca once said, 'The arts are the glue of civilization.' She believes that the cities that have lasted for centuries – Rome and Paris, for instance – endure because they are spiritual and cultural centres. It is something of that legacy that she is trying to create for Baltimore, the city of Poe and TV's *Homicide* and director John Waters' cracked celluloid vision. The challenge is huge. I can't think of anyone else who could take it on.

Gloria Steinem

Gloria is as big an influence on me now as she ever was and she's as inspiring in person as she is in print. Given her impact on my life, it's odd that I can't remember when I first became aware of her. Perhaps it was the first time someone repeated her famous rallying cry for feminists: 'A woman needs a man like a fish needs a bicycle.' Social activism *and* humour – no wonder she was irresistible to young women like me, experiencing the liberating energies of the early 1960s. Gloria found the words for us then and she has gone on doing so ever since as a passionate advocate of women's rights.

A feminist is usually defined as anyone, male or female, who believes in full social, economic and political equality between the sexes. Gloria adds that feminism is about giving each other the power to make decisions. That formula for freedom can be applied to everything from the global politics of the business world to the personal politics of human intimacy. It isn't just about women. Gloria has written and spoken with great insight about the need to raise boys to be nurturing, to let them know that the so-called 'feminine' traits such as compassion, vulnerability and empathy are not things to be suppressed. These are core self-esteem issues, and that makes them very close to the heart of The Body Shop.

But I'm pleased that Gloria also makes a point that I have always felt was critical for our self-esteem campaign. Women are experts at treating other people as they would like to be treated. What they need to do is reverse this golden rule and treat *themselves* as well as they treat others. Passivity is *not* feminine.

No one knows better than Gloria the challenges that face women who stand up for self-determination. So we can take heart from her optimism while we heed her advice: 'Nothing will happen automatically. Change depends on what you and I do every day.'

FALLING APART AT THE SEAMS

The first few years of our move into this land of heroes were extraordinarily successful. Our US sales grew by 47 per cent to $44.6 million in 1993 and profits were up by 63 per cent to $1.9 million. I look back on those early days in America as many of the franchisees themselves have referred to it – as the 'golden age' of The Body Shop. As one put it, 'The sense of conviction, excitement, and fun was palpable.' To me it was a time of some of the deepest friendships. We truly felt ourselves to be merchants of vision.

In the autumn of 1993 we opened our new US headquarters in Raleigh, North Carolina, where we planned to manufacture and distribute US product lines and keep to our goal of having 500 stores in the US by the end of the decade. It was a fabulous area, warm and lovely, where kids could walk to school and housing was cheap. It was a community, like an old American township. There were plenty of people in surrounding towns for the workforce and it should have been perfect. But it wasn't somewhere juicy, like San Francisco, where you get the best marketing ideas, the best of the counter culture. We had set up in a beautiful retreat when we should have been right in the centre of San Francisco or in New York, where we could have got an edge on what was going on. That was a major mistake.

Of course we didn't realize it at first. The area was great for manufacturing and distribution and we built this huge factory, the most environmentally-friendly building in the area, anticipating expansion of between 30 and 40 per cent a year. We should have done a projection plan based on 'What would happen if...?', but we didn't. Then, about a year after the factory opened, the growth of

The Body Shop in the US suddenly slowed. Then it stopped altogether. We were left with huge overheads that seemed to make no sense.

By 1994, our poor performance was attracting media attention. All the pundits were gloating and predicting that we would have to toe the American line, follow the lead of our competitors and become much more of a straight-down-the-line all-American operation with none of this nonsense about campaigning for civil rights and all the rest of it. At the same time, a ludicrous trumped up controversy about our ethical standards hit us (*see* Chapter 10).

As it became clear that our American operation was in difficulties, there was an increasing sense of disquiet among franchise-holders. Competition was increasing from companies which had no interest in ethical standards of trading and were therefore able to rush new products onto the market much more quickly than we could. **Franchise-holders began to complain about being 'burdened' by our ethics.** This was about the last thing I wanted to hear, so I flew to the United States to address the franchise-holders. When I look back now I realize that what I said was both an analysis of our problems in the United States and a résumé of my own curious role within the company.

WHAT I TOLD THE MEETING

'We've heard increasing rumblings of disquiet, with one quarter of you in this room telling us that you're not sure whether your future lies with The Body Shop, and you want to go – some of you faster than others, and some of you with higher expectations of the value of your business,' I said, looking around the room at the franchise-holders, some of whom were among my closest friends.

'Some of you are just tired with the hard work and want time out to reinvent yourselves. Then there are others who want to stay and are desperate to develop but cannot see how to with the territory given. So I can't stand here and pretend to ignore this *Zeitgeist* of The Body Shop America. Every time I come over I'm recognizing increasing symptoms of a general malaise. It's time to change that.'

Change isn't unique to The Body Shop; change has taken over every company. The difference is how you create, master and survive it all.

Often when you read about change, it comes across as a remarkably simple activity: establish your vision, design the change programme and paint by the numbers. Get this: change doesn't work that way. In the real world of change, the vision gets blurred – especially when new leaders come into play. Your staunchest friends cut and run, leaving you with no allies to share your hopes and fears with. Competition and opposition come in the places and forms you least expect, and your fiercest opponent can turn into your most vital supporter. Why does this happen? Because change is all about *people* and people will always surprise you. When you try to bring huge changes into a sleepy business setting, you're going to have some nightmares along with a few sweet dreams come true.

I told them I saw the problems, issues and questions like this:
- ☛ There's been a stultifying lack of creativity and an obsession with looking at the competition, yet creativity is the only way to set us apart.
- ☛ We've seen major success in product development – but what about human development?
- ☛ We've seen some new product ideas – but where's the marketing differentiation?
- ☛ We've seen the brave expression of our values in the Ogoni Freedom Center in Boulder and New York – but what of the day-to-day values of caring for each other?
- ☛ We've had lots of activity in the alternative press and among NGOs – but there has been very little positive coverage in the mainstream media.

'What we need to do now is to see how we can revitalize our company and how we can renew people with an emotional energy,' I continued.

'In the last decade we've presumed so much: that if we opened our doors, the customers would come in; that all we needed was more shops in more malls; that having a vision of great values actually leads to acting out of values; that we had something unique and valuable, when in reality we've been fodder for the competition. The last thing we must do is force change on any of you, because if we inflict change, then you may stay with us when you've mentally quit; you may say 'yes', but do 'no'. What we need is to create the right environment – one where you feel part of the change and where you feel comfortable going in the direction we are leading you towards. We have to beware the trap of rationality: just because something looks 'right' on paper doesn't mean it's the right thing to do in our company.'

THE NADIR

That meeting with the franchise-holders was rock bottom in the US. Looking around the room at the time, I saw only two options: complete reorganization or closing down in America. But in fact it was the beginning of an enormous re-engineering, restructuring and rebuilding of the US company, something we had no track record of doing before. Ironically, even at that point, our brand was becoming increasingly recognized across the USA, but the other difficulties were real.

I could not blame the people who ran The Body Shop stores for being concerned. By 1994, we were experiencing, for the first time, people choosing to no longer shop at The Body Shop. For every 100 customers, we were losing five. The stock market decided we were no longer their darling and hacked our share price to pieces, while the media rekindled the feeling of the 1980s and began calling for us to step in line, grow up, behave and go the way of other large corporations.

Ironically, our greatest achievement in the United States appeared to have been to show the way for competitors. Copycats, small and large, proliferated everywhere. We appeared, quite simply, to have given them a blueprint. And while our legal department – God knows how many lawyers we employed – could protect our name, our logo and our designs, there was absolutely nothing anyone could do to stop lookalikes of The Body Shop opening everywhere.

We didn't understand it was a different culture out there. We

The fact is, we just got everything wrong.

didn't understand that price promotions made it virtually impossible to know the real price of anything. We didn't understand how much retailing depended on advertising. We had dull-looking shops and the wrong products. We thought we ought to put Americans in charge, but too often we chose the wrong Americans. We were arrogant. We didn't change fast enough – we had the same shop look for 10 years in a country that is retailing heaven. Americans want everything new and they want it now – instant gratification, with little ethical or emotional baggage to clog up the system.

For a fresh view we brought in non-executive directors who said that we had to have better marketing, but then we just didn't listen. We were holding extraordinary franchise meetings where there was a huge sense of family and fun and we were so proud of our campaigns that we forgot to watch the little mule that was bringing us in the money – we didn't give him enough food. We didn't take the time to stop and reflect on what we were doing; we were being driven by hurry sickness and it was driving us to the brink. We should have said: 'No more mall shops.' Our place was in the cities, the urban areas, on

think my mistake in the United States vas that I didn't interfere enough.

main streets, places that were fun and idiosyncratic. We never understood the malls or the mall mentality.

I was furious that we were constantly being compared with Bath & Body Works, when there is simply no common ground in either attitude or objective. I am much more interested in what Aveda or Origins are doing, and some of the more idiosyncratic groups. Bath & Body Works are just another fragrance and lotion company. If they are making more money, well, money has never been my primary objective. I heard that the chairman of Bath & Body Works had a sign on his wall which said: 'Bury The Body Shop'. That is a mission statement with absolutely no inspiration or imagination – and not much staying power either.

Some people might say I interfere too much, but in America I let myself be persuaded by others. We didn't have any space or time for experimenting, for fun. I wanted to paint our shop in San Francisco pink, just for fun, just to see what would happen, but we never got round to it. When sales began falling we brought in a whole group of marketing people and product development people and they just looked around at what was going on in the market place and said we had to do the same. I should have stepped in and told them that if that's the way the market is going, we should go in the opposite direction.

AMERICAN REVIVAL

What we should have done from the start was to remain different. We should have adopted a micro-market approach and opened dual language stores – Spanish and English in California and Japanese and English in Hawaii. We shouldn't have gone into shopping malls. If you

were to drive across America stopping at every mall you would find each one very much like the one you had just left – the very antithesis of what we want for The Body Shop. **We should be urban, wild and eccentric.** We shouldn't have gone down to North Carolina to set up our operation. The bigger we get, the more local we should be. In the future we should be winning awards for micro-marketing.

But we've already begun to redeem ourselves in the United States, by being wild, even feral, by being like ditch weed, growing in unexpected places. We completely redesigned and re-energized the business in a process that took about 18 months. The Body

Shop is doing very well in the cities, though less well in the provinces, where shopping malls dominate the retail market.

Looking back on these years of the 'easy life' in America, I think of how the franchisees became our friends and we, as a group, were brave and indulgent. We brought community activism into the shops. Employees were doing stints in a Romanian orphanage, visiting our community trade initiatives, being political. PhD students were writing dissertations on the values of The Body Shop. I look back with affection on that half a decade in America and still believe it was the time of the greatest generosity of spirit.

We would never have got where we are if we hadn't had this absolute plethora of shops and franchisees, and that is wonderful when you want to build a brand, or a company.

So The Body Shop will remain in the United States and you won't find us copying any of the dozens of companies that have copied us. I hope to God we will always be the leader – if not in profits, then in ideas and principles. In those areas, I don't believe anybody can offer us any competition.

THE OGONI CAUSE HAS CAPTIVATED OUR COMPANY. WE MADE A PROMISE THAT WE WOULD SEE THIS CAMPAIGN THROUGH AND USE EVERY OPPORTUNITY TO PROMOTE THE OGONI MESSAGE.

WE ARE GOING TO KEEP THAT PROMISE ...

KICKING UP A FUSS ON THE CAMPAIGN TRAIL

I first became aware of the plight of the Ogoni people through the Unrepresented Nations and Peoples Organization, which we had been helping to fund since 1991. Little did I know how much sheer energy and passion we would be investing in the campaign for them.

The Ogoni have lived in the Niger Delta for centuries. Mainly farmers and fishermen, they enjoyed a tranquil existence largely undisturbed until Shell discovered oil in the delta in 1958. Then, in a few short years, their tribal homelands were ravaged and they became embroiled in events far beyond their control. Oil quickly became Nigeria's economic lifeblood, accounting for some 80 per cent of government revenue. It was when the Nigerian government acquired powers to take over tribal land for further oil exploration that the gentle Ogoni were virtually doomed. Largely illiterate and unaware of what was happening until it was too late, they stood by helplessly as contractors cleared their farms and destroyed their crops. Soon the Ogoni homeland was scarred by pipelines, gas flares and oil waste.

We have woken up,' one of them wrote, 'to ind our lands devastated by agents of eath called oil companies. Our atmosphere as been totally polluted, our lands

Notes for my speech about Ken Saro-Wiwa, May 1995

degraded, our waters contaminated, our trees poisoned, so much of our flora and fauna have virtually disappeared.'

Marginalized by economic and political considerations, like so many other indigenous peoples, the Ogoni reaped no benefits from the vast sums paid by the oil companies to the Nigerian government – most of which was spent on military hardware or spirited out of the country to be deposited in numbered accounts in Swiss banks.

In the early 1990s, the tribe decided to fight back and began a campaign for environmental, social and economic justice by demanding political autonomy. In June 1993, I met a delegation of Ogoni activists at the United Nations International Human Rights conference in Vienna and I was deeply impressed by them. Their outward gentleness was in stark contrast to the violent story they had to tell of how their homeland had been plundered and how their farms had been razed by Shell contractors operating under the protection of government security forces. I learned how peaceful demonstrations against the construction of pipelines had been met with brutality and how their leader, writer Ken Saro-Wiwa, had been arrested on his way to the conference.

For Gordon and me, the Ogoni became a powerful symbol of big business running roughshod over the interests of small people and I had no doubts that we should do everything in our power to help them. We never felt any anxiety about going up against a company as big as Shell, even though everybody else seemed to be nervous about it. **When you do things instinctively, you don't think it's frightening or remarkable, you just know it's the right thing to do and that power pushes you on.** But when we approached other companies with known records of fighting for human rights, asking them to help us on this one, none of them would.

The first thing we did was launch a letter-writing campaign to demand Ken's release. While that was going on, we provided the Ogoni Community Association in London with office equipment to help them organize and paid for a group to fly from Nigeria to Geneva to address the UN Subcommission on Human Rights.

Few of the Ogoni had ever heard of The Body Shop – why should they? But when they became aware of our involvement in their fight they were astonished to discover we were only a skin and hair care company. I remember one of the people who had travelled to Geneva at our expense telling me how surprised he was when he went out for a stroll one night, came across a branch of The Body Shop and realized that this was the company which had made it possible for him to be there.

We did everything we could to resolve the problem behind the scenes. Gordon went two or three times to Shell to talk to them about how the situation could be worked out. We told them they didn't have to do what they were doing, that we would pay for a team of experts to go down to the delta, assess the environmental and social conditions and offer alternative solutions. They gave us a polite no. Their argument was always that they had no control over what was happening, that it was down to the Nigerian government.

They simply couldn't understand why a company like The Body Shop would get involved in their business. Their attitude was more or less: 'What has this got to do with you?'

In August 1993, Ken Saro-Wiwa was unexpectedly released from prison. He flew to the UK and came to Littlehampton to thank us for what we were doing. Unfortunately I was away at the time, but he made an enormous impression on Gordon and everyone else he met. It reassured us that what we were doing was right.

The following year, at the invitation of the Ogoni, we planned to visit Nigeria to carry out an environmental impact study of Shell's operations in the area, but all our people were refused visas. We were bitterly disappointed and desperately frustrated,

Where there is one slave there are always two – he who wears the chain and he who rivets it. Jeanne de Hericourt since the situation in the Niger Delta was deteriorating rapidly. Many Ogoni were being killed in 'ethnic clashes' with 'neighbouring communities', yet their opponents were using sophisticated weaponry and mortar grenades, which we very much doubted were available to simple delta fisherfolk. Later the NGO Human Rights Watch would collate evidence from Nigerian soldiers who admitted they had been involved in attacks on Ogoni villages. No doubt Shell's executives were appalled, but it was the consequence of a large and insensitive multinational being prepared to do business with a brutal tyranny.

The Nigerian government introduced the death sentence for anyone found guilty by special tribunal of 'involvement in community clashes'. A few days later four conservative Ogoni chiefs were lynched when they were suspected of collaborating with the military authorities. Despite the fact that Ken Saro-Wiwa was nowhere near the scene of the killings, he and a number of his supporters were arrested and charged with murder. From that moment on, the Nigerian military regime instituted a reign of terror in Ogoniland under the pretext of establishing law and order in the area. Some 50 Ogonis were killed by security forces and more than 600 were arrested, thrown into jail, held without trial and tortured during interrogation.

At the same time, The Body Shop campaign was escalating very rapidly. We put posters in our shop windows and on our trucks. We staged demonstrations, organized cultural events, advised the Ogoni, talked to the press, lobbied politicians and waved banners outside Shell's headquarters in London. The Body Shop staff world-wide telephoned the Nigerian government in Port Harcourt to protest against Ken's continuing detention and many of the departments at Littlehampton set their fax machines on permanent 'send' in order to bombard the Nigerian High Commission in London with letters. When the Commission eventually turned its machines off, we hand-delivered thousands of letters we had collected.

Such was the strength of emotions surrounding the plight of the Ogoni that many individual shops staged their own protests.

Aaron Battista, who was the regional sales manager in New York, saw the Ogoni action station set up in a warehouse at Littlehampton and realized he could do something similar with unused space above The Body Shop on Fifth Avenue in Manhattan. I thought it was a brilliant idea and so he set up the first Ogoni Freedom Centre, which attracted dozens of visitors every day, along with school parties. They got a comprehensive overview of the oppression of the Ogoni by Nigeria's military regime acting in cahoots with global oil interests. The centre also offered the opportunity to write, fax or phone Shell, the Nigerian Embassy and President Clinton. To me, this was an example of real people power and was what set us apart from other businesses.

In May 1994, we picketed the World Petroleum Congress in Stavanger in Norway. We set up a mock 'welcome desk' at the airport and gave arriving delegates a goodie bag containing The Body Shop products and protest letters. There was no question of us stepping down.

Not long afterwards we received a letter from Ken that had been smuggled out of prison in Port Harcourt:

I am fully aware of all that you have done for the Ogoni cause and will thank you forever for it... The Ogoni are down and out and, but for The Body Shop, might well have gone under. You give us a lot of hope. I pray that you succeed in keeping the momentum going.

While we were campaigning, the world's governments were apparently engaged in what they like to call 'quiet diplomacy' – and it was completely useless. To our horror, in October 1995, Ken and eight other Ogoni activists were sentenced to death. We did everything in our power during the next few weeks to alert the world to what was happening in the hope that international pressure might save them, but to no avail. On 10 November 1995, Ken Saro-Wiwa and his fellow prisoners were brutally executed.

For me personally and throughout The Body Shop there was an overwhelming sense of grief and despair that we had been unable to save them. But Ken's dreams of freedom for the Ogoni

In Nigeria it's easier to bury the protesters than the oil pipes.

My newsletter to The Body Shop magazine *Gobsmack!*

people were not in vain because the fight against the Nigerian military government continued. In the wave of outrage that swept the world after the executions, Nigeria was suspended from the Commonwealth, but we did not think it was enough. We appointed a full-time human rights campaigner to co-ordinate the campaign and were able to help some of Ken's family escape from Nigeria.

We made sure the situation in Ogoniland featured prominently in all our publications:

On 4 January 1996, Lucky Garabe, a 12-year-old boy from Bo-ue, Ogoniland came out onto the streets of his village to mourn Ken Saro-Wiwa's death and to celebrate Ogoni Day. He was just one of over 100,000 other Ogoni women, children and men who danced and sang in Ogoniland. Wherever they came out, they faced well-armed security forces. The soldiers watched at first and then the bullets began to fly. One of those bullets hit Lucky in the back of his head. He died instantly. Two days passed before his body, which was taken by the soldiers, was found floating in the Ko river.

Where did Lucky and those thousands of others find the courage? It's a simple answer; one that Ken Saro-Wiwa died for. They found their courage in ideas. Simple ideas, like the right to a clean environment, the right to piped drinking water and the right to a say in their own future. The international campaign in support of the Ogoni people's demands has also given them courage. Which is why The Body Shop, with so many other human rights, environmental and pro-democracy organizations, was committed to continuing in 1996.

The oil companies operating in Nigeria, led by Shell, are sure to face an internationally co-ordinated boycott, one to which The Body Shop is already committed. Meanwhile the Nigerian military dictatorship may look isolated by the Commonwealth's unprecedented suspension and sporting bans but we still have a real fight to keep the pressure on. Especially when a multi-billion dollar business deals with the murderers who rule Nigeria only days after Ken's execution. The Ogoni have the courage to march and we will continue to march with them.

After Ken's death, we focused our attention on winning the release of the 19 remaining Ogoni activists still held in detention. I'll never forget the letter they wrote to us, addressed to 'Mr and Mrs Bodyshop', which was somehow smuggled out of the prison where they were being held. They first described the conditions they had to endure:

We are held incommunicado in an overcrowded and ill-ventilated cell measuring 20 by 28 metres with more than 120 inmates sleeping in three rows along the length of the cell on tick, lice and bedbug-ridden mats. We are only allowed to take our bath just twice a week from a well which was recently a dumping pit for dead inmates and still contains human skeletons. This is also the source of our drinking water...

They then continued with a heartfelt plea for us not to forget them and expressed their thanks for what we had done:

Mr and Mrs Bodyshop (as we fondly call you), it will not reckon well with our conscience as it will amount to gross ingratitude if we should fail to appreciate and acknowledge your enormous assistance to us since this tempestuous period of the Ogoni people. In short, your names are synonymous with the Ogoni struggle for survival and by this you have engraved your names not only in the annals of the Ogoni people but also of all the ethnic minorities the world over. We cannot find the appropriate word to express our profound gratitude and appreciation to you but we fervently pray God to grant you the fortitude and wisdom to continue to render these selfless services to humanity...

When Gordon replied, we heard that they were so overjoyed to learn that their letter had reached us that they began singing and dancing in their cell, much to the astonishment of the prison guards.

The 'Free the Ogoni 19' campaign was supported in 17 of The Body Shop markets from Japan to Germany. In Ireland that Christmas, our shops displayed posters featuring the Ogoni 19 instead of traditional Christmas stuff. Can you imagine any other business not promoting its products at Christmas?

Our campaign resulted in hundreds of thousands of postcards and letters being sent to Shell and the Nigerian government, which seemed as bewildered as Shell by our involvement. We acquired a copy of the minutes taken at a meeting between Shell executives and the Nigerian High Commissioner in London in which he expressed dismay about 'the unusually strong commitment which Mr and Mrs Roddick have given to Ken Saro-Wiwa's case' and protested about the 'apparent network of misinformation orchestrated by Gordon and Anita Roddick ... against the trial of Mr Ken Saro-Wiwa...' **At that meeting Shell was apparently forced to explain who the hell we were.** When shareholders arrived for Shell's annual general meeting, they found the offices besieged by people carrying placards produced in The Body Shop design department, with the faces of the Ogoni 19. Staff throughout our shops world-wide 'adopted' members of the Ogoni and struck up friendships with their families, writing them letters and sending them messages of hope from around the globe. Together with the cash provided by The Body Shop Foundation, the charitable arm of the company, this made an enormous difference to the welfare and morale of the whole Ogoni community.

We realized that Shell was shaken by the fiasco when the UK chairman and chief executive, Chris Fay, told a group of activists and politicians in London:

We have to consider why trust in companies is declining. I think that the roots of this mistrust lie in the fact that people increasingly fail to see the relationship between business success and their own quality of life... They are suspicious that business standards do not protect people and the environment ... and don't understand how business can contribute to achieving a sustainable future.

In March 1997, as a response to the storm of criticism about its activities in Nigeria, Shell issued a revised operating charter that committed it to human rights and sustainable development as an integral part of the company's policy.

I was sick at heart that it took the lives of Ken Saro-Wiwa and who knows how many more Ogoni to get Shell to concede that it must become more transparent in its operations, but it was a definite

Rhetoric never won a revolution yet.

Shirley Chisholm

step forward that a leading multinational corporation should at last define for itself a more legitimate role.

The Ogoni 19 – actually by then the Ogoni 20 – were finally released unconditionally in September 1998. After four years in prison, their continued detention was ruled 'unconstitutional, unlawful, illegal, null and void'. It was fantastic news. The same year, Shell launched its 'Profits and Principles' advertising campaign, which it claimed was 'the first external sign of our renewed commitment to recognize the legitimate interests of a much wider group of stakeholders in our business and our need to listen, engage and respond to them'. I suppose we could claim this as a partial success, but I still grieve for Ken Saro-Wiwa and his like.

What on Earth is a skin and haircare company doing getting involved in political activism anyway? Why don't we just shut up and stick to selling shampoo and soap? Well, one good reason – at least for me – is that I am not interested in *just* selling shampoo and soap. The simple fact is that I'd rather promote human rights than a bubble bath. The second reason: if not us, who else? The alternative (maybe worse): doing nothing. **Quite apart from anything else, my experience is trying to change things for the better makes you feel better, healthier. Humans are communicative animals: when you do good in a community, the benefits eventually get back to you.** I also think the business community has tried to operate politics and commerce in completely separate arenas for far too long, believing that neither should interfere with the other. I disagree fundamentally – I'm for interference. As far as I'm concerned, political awareness and activism must be woven into the fabric of business. In a global world, there are no value-free or politically disentangled actions. The very act of organizing on a global basis is political because of culture, geography and differing value systems.

But then, I think activists – whether for human, animal or environmental rights – are the real heroes and heroines of our society. They can stand up for positive grassroots change in ways that most people in business or politics can only look upon in awe.

The Body Shop recognized long ago that politics is far too important to leave to politicians.

So we dedicated our business to the pursuit of human rights, of positive social and environmental change, and got on with becoming an activist organization pursuing those objectives. That meant action, not just words.

Most of the people I have worked with have these activist yearnings. For them, work is about a search for a daily meaning as well as their daily bread, for recognition as well as cash, for astonishment rather than torpor. I think most people join The Body Shop because they want to be part of the movement for social change. We're a skin and hair company that works for positive social change. We stand by our products for efficacy. We stand by our ethos for advocacy, and our employees actively join in. They're part of what we stand for – the real in an industry selling the impossible. That gives us an automatic political stance, one dedicated to honest dealing, supporting human rights and protecting animals and the environment. As a business, that puts us in a unique place.

None of our competitors have followed our example because it doesn't create sales – believe me, if it did everyone would be doing it. They can copy our products and our style and pretty well everything else, but they will never follow us into the arena of political campaigning. So I can't say that the social activism programme of The Body Shop benefits our financial bottom line. But what it does do is give us an identity that is recognizable. It provides motivation for our employees and sets us apart in a way that I think that is more engaging and more meaningful than any advertising campaign.

Campaigns such as ours seem to work best against companies that depend on strong consumer brand loyalty and less well when there is less brand awareness. Campaigns that focus on child labour or sweatshops or the environment appeal more to American consumers than those about union rights. Success also depends on developing a monitoring capacity in country after country to hold companies like Starbucks, Gap or Nike to their promises. I'm always disheartened by the fact that most media outlets are controlled by large corporates and thus make it more difficult to get any story about citizen campaigning – let alone a positive one – into the press.

MY FIRST SENSE OF OUTRAGE

My personal capacity for moral outrage was stimulated when I was about 10 years old and I picked up a paperback book – one of those cheap editions that were just starting to get popular in the 1950s. It was about the Holocaust. There were six pages of photographs from Auschwitz and they made such an impression on me that I can describe every one of them to this day. Seeing those pictures, I experienced such a sense of injustice that it shaped the kind of person I became.

From that day onwards I became a shouter, a marcher, a teenage campaigner. I drove my mother mad, pestering her for subscriptions to whatever good cause I had just stumbled across. I was also lucky in that I had a series of teachers who never tried to suppress my enthusiasm. They knew that the only books I was interested in were books by social writers – certainly not Henry Miller at that stage, but definitely Steinbeck and Faulkner – so my entire literary education consisted of the American and English social writers of the 1930s. My teachers never tried to force me into a curriculum, instead they allowed me to develop the areas that I was interested in, and I thank them for it, because I am sure that they put in place the planks of that platform of social outrage. They legitimized everything; they said it was all right for a 13-year-old to go on vigils in support of the Campaign for Freedom against Hunger or march with the Campaign for Nuclear Disarmament. Nobody said, 'Oh, don't be stupid, you should be playing with your Barbie dolls.'

Fortunately, Gordon also has a highly developed social conscience. When we first got together we were both involved in social and political issues in our local community in Littlehampton. We worked with Shelter, helped the homeless by offering them accommodation in our hotel out of season and joined the Squatters' Association, which encouraged the homeless to occupy empty houses. The blackboard in our restaurant usually carried political or social messages rather than a menu.

I can't honestly say when I opened the first branch of The Body Shop in 1976 that I had any inkling of what we were starting – I was a bit more concerned about making enough money to pay the bills and stay afloat. In fact it wasn't until after we had gone public in 1984 that it began to dawn on Gordon and me that – no matter how corny it might sound – The Body Shop actually had the potential and power to do good.

CAMPAIGNING ...

WITH GREENPEACE

The notion of harnessing commercial success to altruistic ideals set my imagination on fire. From that moment on, The Body Shop became a company with *attitude*. We made our campaigning debut in 1985 when we teamed up with Greenpeace – which had never joined forces with any kind of commercial enterprise before – to demand an end to the dumping of poisonous waste materials in the North Sea. The Body Shop paid for 100 giant posters to be displayed on huge advertising hoardings, showing a Greenpeace ship fighting the waves with the slogan 'Thank God someone is making waves' and a small note saying people could join Greenpeace at their nearest branch of The Body Shop – the only indication of our involvement.

This led naturally to another joint campaign the following year to stop the slaughter of sperm whales. These glorious creatures were being threatened with extinction – their numbers had dropped from more than one million to less than 50,000. Whales were particularly significant, as the cosmetics industry used the ambergris oil from them in many beauty products.

There were rumblings from some staff and a few franchise-holders that we were becoming 'too political' and getting into areas that had little or nothing to do with the business, but by then I was absolutely committed to the idea of The Body Shop as a force for social change. How else could one add value to a skin and haircare company? No one *needs* anything we sell. You can rub mayonnaise into your hair to condition it and you can use salt to soften your skin. **None of our products are a matter of life and death, so campaigning for me became the method by which we could introduce values into a non-value industry.**

WITH FRIENDS OF THE EARTH

I wanted to extend the 'Save the Whales' campaign to our shops abroad but ran foul of Greenpeace's bureaucracy. They had to get permission from their individual branches in each country and none of them seemed very keen. I knew that some people in Greenpeace were unhappy with the idea of joining forces with us, but I was amazed that an NGO would be so hidebound. I began to search around for a new buddy and mentor who would trade knowledge and information in return for campaign support. Friends of the

Earth came top of the bill and appeared to have no reservations about working with us.

Unfortunately our first campaign with Friends of the Earth – against acid rain – was a disaster. We hired a Polish designer to produce a very quirky surreal poster with a dead tree sprouting from a decomposing human head against an industrial background of smoking chimneys. The copy line said simply: 'Acid Reign'. We thought it was very clever, but unfortunately nobody else did. In fact no one seemed to know what we were on about. I insisted that the campaign went into every branch of The Body Shop world-wide, but instead of tackling acid rain pollution we really only mystified our customers.

Even so, we learned a lot from that experience about the need for simple, emotive imagery and went on to forge a very successful partnership with Friends of The Earth, campaigning for protection of the countryside, against CFCs and to raise awareness on a whole range of environmental issues world-wide.

Later we joined forces with any number of environmental and human rights organizations, among them Amnesty International and Survival International.

AGAINST ANIMAL TESTING

Our campaign to ban animal testing in the cosmetics industry came up against the deeply ingrained vested interests of the big players in the industry. When in the late 1980s the EEC threatened to bring in legislation that would actually make animal testing compulsory, it threatened the very cornerstone of our business. At the time I said I would close The Body Shop down rather than comply – and I meant it. For more than two decades we had been shouting about the evils of animal testing, but just when we thought we had got it right and Brussels was finally going to ban products that were tested on animals, the big religion – free trade – raised objections. We collected four million signatures in a single year to protest against animal testing and in November 1998 the United Kingdom at last introduced a ban on animal testing for cosmetics.

Why test on poor defenseless little animals when they could use my husband?

WITH THE BODY SHOP FOUNDATION

In 1990, we established The Body Shop Foundation as a vehicle for the distribution of our charitable donations. The Foundation's resources are directed towards supporting those individuals and organizations working in the fields of human and civil rights, animal protection and environmental protection. Its mandate is to 'effectively take risks', which means it is particularly interested in working with small innovative groups and activists experimenting with new ideas. In its first six years of operation, the Foundation donated more than £3.5 million to more than 180 charitable groups around the world. But it isn't just a 'cheque-book charity' — it also offers support in kind, in expertise, commitment and networking. That's part of our campaign work too.

IN ROMANIA

Sometimes you have to go over politicians' heads and go straight for the public's heartstrings if you're going to get things done. It achieves far more, especially when politicians fail as miserably as they have in Romania, Albania, the Balkans and Bosnia. I've been there and seen for myself.

The same year that we set up The Body Shop Foundation, I made my first trip to Romania, visiting one orphanage after another full of abandoned, unloved, unwashed and unstimulated children. I was absolutely appalled. Back in Littlehampton, I set about recruiting a volunteer team to go out to Halaucesti, a little Moldavian village in a far northern corner of the country. The Romanian Relief Drive, now called Children on the Edge, was up and running within six weeks. We set about refurbishing three orphanages, helped set up a permanent healthcare team and funded the testing of more than 4,000 children for disease and infection.

By now volunteers from The Body Shop have worked with orphaned children in Romania for fifteen years. Our long-term

Resistance to tyranny goes by the name of protection.

Emma Goldman

commitment is to refurbish buildings and provide comfortable and sanitary living conditions for orphans, along with healthcare, in order to provide a model for institutional care in Romania and change the existing system, which succeeds only in isolating and depriving children of the support and love they need. We want to give the children their childhood back. Later we extended our volunteer work to Albania, East Timor, Kosovo, Moldova and Bosnia.

WITH 'MAKE YOUR MARK'

To mark the fiftieth anniversary of the Universal Declaration of Human Rights in 1998, The Body Shop in partnership with Amnesty International undertook the 'Make your Mark' campaign, the largest human rights campaign in the company's history, which was launched with the support of His Holiness the

Dalai Lama. We collected over three million thumbprints in 34 countries in support of 12 remarkable human rights campaigners who defend fundamental human rights, often in dangerous and threatening conditions. Each country then selected a local artist to create a portrait of their adopted defender using the thumbprints.

People often ask if our campaigns are effective and whether we actually achieve anything. I think we do. When we worked with Amnesty International, the leverage we were able to exert through our shops generated so many letters that 17 of the 30 prisoners of conscience we were allocated were released. That's brilliant, but what we really do well, all over the world, is raise awareness. The Australian shops, for example, really brought the question of French nuclear testing in the Pacific to public attention with a fabulous campaign called 'If it's so safe, do it in Paris'.

Never doubt that a small group of thoughtful, committed citizens can change the world; indeed, it is the only thing that ever has.

Margaret Mead, anthropol

By campaigning to end animal testing in the cosmetics industry, we caused cosmetics companies to change their practices and put a moral issue on the political agenda. We achieved this by practising what we preach, finding common cause with those who shared our vision and – above all – channelling the energy of our employees, franchisees and customers. The same passion has guided our progress on the struggles of indigenous peoples to move from the shadows of political and economic marginalization. Our action on violence against women confronted prejudice, rebuilt personal self-respect and forced politicians to confront the impact of their actions, or lack of them. Through our actions we have helped marginalized communities locally, nationally and internationally to find a voice. And our steadfast support for Ken Saro-Wiwa and the Ogoni people of Nigeria forced a transnational corporation to confront its environmental and human impact and to defend its actions before the public, politicians, media and – even more important – itself.

As many as 67 per cent of shoppers take a company's ethical position into account before buying one of its products. As a conscientious consumer, I would be thrilled to know that the money I spend goes towards issues like human rights, changing the nature of international trade, celebrating women, setting up initiatives like *The Big Issue* and supporting groups like Missing Persons Helpline. I would be thrilled to know that a company isn't just mouthing slogans like 'Against Animal Testing', but is actually trying to change the law. I would be thrilled to know that the people who serve me spent their sabbatical time working in Romanian orphanages or Albanian mental institutions.

People don't want simply to buy the product, they want to have sympathy with the company too.

CAMPAIGNING STAFF

When people come into The Body Shop to sign a petition they don't say: 'Oh, by the way, I want some moisturizer.' That is why most companies practise chequebook charity rather than campaigning and activism. Bringing activism into the store confuses people, but it certainly inspires your staff.

The Body Shop employees are pretty dedicated and we've got a really good network. The campaign has to be inspiring to the employees because if it was just an add-on, I would tell them to stop. And the truth is, without inspiration it probably wouldn't work anyway. One of the key elements of successful campaigning is the energy and enthusiasm of the staff. They are the front line, the publicists for any campaign, because they are the people who interact with the customers and the people who take part in working out and acting out The Body Shop ethos.

Over the years our employees have become very skilled at campaigning. They have learned to chat up the local newspapers and radio stations and come up with brilliant ideas on how to become engaged. They can see how, through the platform of the shop, they can really make a difference.

Several years ago the young people running stores in Edinburgh became the forefront of the Ogoni lobby at the Commonwealth Heads of Government conference, demonstrating, marching with nooses and sweatshirts, and dedicating their stores to keeping up the pressure to free the Ogoni 19. During the 'Make your Mark' campaign, people were doing the most extraordinarily imaginative pieces of street theatre, turning the shops into prisons and other simply amazing things. We give them the notion and they interpret it and add to it, because of the kind of people they are – creative, entrepreneurial individuals, passionate in their belief in The Body Shop. I discovered very early on that there are few motivational forces more potent than giving staff an opportunity to exercise and express their idealism.

Let me share another example with you. When a member of staff looks you dead in the eye after three exhaustive weeks refurbish-

The realization that working for The Body Shop involves more than just stocking shelves and selling bars of soap generates an enormous sense of pride and commitment.

ing a Romanian orphanage, holding babies with AIDS or campaigning for human rights, and says, 'This is the real me,' take heed, for she is dreaming of noble purposes, not a moisturizer.

I believe that giving staff the opportunity to express their idealism is one of the moral obligations of businesses. If you want to create a sense of spirituality by service, then you've got to encourage volunteering. Our employees want to change the way things are, they want to do something for the common good, and they come back charged with a feeling of commitment and a sense of community.

As the majority of our customers and staff are women, we feel a special responsibility to campaign on women's issues, especially on domestic violence against women. In 1995 we galvanized 17,000 UK customers to send postcards to the government to get the Family Law Bill altered so that unmarried women had the same legal rights and status as married women when they needed to defend themselves against domestic violence. Our customers sent so many postcards to the Lord Chancellor that he couldn't open the door to his office.

Once the staff are on board, our real power is the ability to demonstrate that millions of ordinary people care passionately about human, social and environmental rights. We have a core of people who see this as part and parcel of shopping at The Body Shop and who are almost disappointed if they go into a shop and find there isn't a petition or a place to write a letter or send an e-mail. People view our shops rather as they might view a community centre or a library – as a place to go for information. We are constantly bombarded by school children, parents and teachers looking for information about various projects. And The Body Shop staff, being who they are, will always try to help them.

Anything that changes values changes behaviour.

There's a great learning lesson there. People's power is then held in dreams, curiosity, music, and a desire to reach for the human spirit. You can find it all through that secret ingredient called enthusiasm. Enthusiasm created from the heart guides the whole person, so there is no resistance, and everything flows and seems possible. If we start feeling positive about politics, great things can happen. Politics is the art of the possible.

The voice of objection has sometimes been obscure, silly or wilful. But more often it has been serious, brave and inspirational.

It was almost a year to the day after the release of the Ogoni 20 that I boarded a plane for Port Harcourt in Rivers State, Nigeria, accompanied by Gordon, our Irish head franchisee Peter MacDonald and other colleagues from The Body Shop. This was my opportunity to experience the Ogoni people, culture and environment.

Even though it was 5 a.m. when we arrived, hundreds of singing and dancing Ogoni welcomed us at the airport. We spent the next week being fêted by thousands more. Gordon and I were honoured by being made Ogoni chieftains and 'Wipers Away of Ogoni Tears' at a ceremony attended by thousands of these remarkable, resilient and resourceful people.

In between the festivities, there was one occasion that meant more to me than anything else. It was meeting the Ogoni 20. The room was hot and crowded and there they were with their own T-shirts saying: 'The Ogoni 20 welcome The Body Shop'. As I sat in that room, I looked at these men and their families and heard what our letters, postcards and protests had meant to them, and at that moment I could both see and feel why campaigning works.

While there was plenty to celebrate on the visit it was also a valuable opportunity to see the injustice of the Ogoni people's poverty. Ogoni villages are only steps away from industrial oil facilities. You can still see oil spills that have devastated the land around them. What

I found most disturbing was the Shell community hospital in Gokanna, which Shell took over from the government earlier this year. Shell claims it supplies the hospital with medicines. In reality, this means Shell sells the hospital medicines but for the patients it is cheaper to go to Port Harcourt and buy their medicines there. The hospital caters for up to 200,000 people and has shockingly inadequate facilities. Yet 30 miles away, there is the Shell employees' hospital in their Port Harcourt compound.It is one of the most modern and best-equipped for treating tropical diseases in the world.

The Gokanna hospital may not be typical of the medical services which Shell supports, but the sad reality is that the company is still not welcome in Ogoniland. The company has issued a social report and run a multi-million dollar advertising campaign stressing its commitment to social and environmental responsibilities, but the demand in the delta for effective engagement leading to reconciliation and the flow of much-needed resources to the Ogoni people still has an extremely long way to go, as far as I'm concerned.

When and if Shell does get it right with the Ogoni, the multinational will have a model to take forward to all the other delta communities on which it has an impact. The Ogoni are a symbol for all the oil-bearing communities across the Niger Delta. Today there are protests spreading across the region, a bushfire of discontent. While the Ogoni have survived, they have seen little benefit for choosing the non-violent path to change. Governments and the oil companies would do well to demonstrate that the non-violent path does bring rewards.

Now that the soldiers are gone and the Ogoni prisoners released it is easy to think the struggle is won for the Ogoni. But behind all the songs of thanks, I heard a plea that went, 'Don't forget us.' While they have survived one struggle, their next has only just begun. We mustn't forget them.

OUR TRADE WITH THESE COMMUNITIES IS NOT JUST ABOUT CREATING ANOTHER PRODUCT OR MARKET FOR THE BODY SHOP.

IT IS ABOUT ...

MERCHANTS OF VISION

It all began in February 1989 at the Forest Peoples' Gathering at Altamira in Brazil, which was organized to protest against proposals to build five dams in the Xingu river. Building these dams would have resulted in the flooding of 10,000 square miles of land which was home to a large number of indigenous people. The threat to the Amazonian rainforest was of immense concern to environmentalists: millions of acres had already been destroyed by logging and the rainforest inhabitants had been thrown off their land, ravaged by diseases against which they had no immunity and used simply as a source of cheap labour. On top of that, the loss of flora and fauna was incalculable – the rainforest provides the sole habitat for half the species on Earth, many of which still wait to be discovered.

I was invited to attend the gathering as I had been involved in campaigning against the destruction of the forest. I decided to go, as it would be an opportunity to talk to many of the world's leading ecologists and environmentalists and to learn about the problems of the rainforest at first hand. It was an extraordinary experience, particularly when many of the forest tribes turned up in full ceremonial regalia, with body paint and head-dresses of yellow, red and green parrot feathers. They had extraordinary presence and many of them spoke with stunning eloquence. They told the world, in no uncertain terms, to leave them and their forest alone.

This was a fabulous, heady, inspiring period when I really thought we could succeed at things no other company had even wanted to try – starting with saving the Amazonian rainforest. But our first, most adventurous and most risky attempt to build trading links with indigenous tribes led us on an incredible journey.

THE HOPE OF TRADE

The most impressive of the speakers at the Forest Peoples' Gathering was Paulhino Paiakan, a leader of the Kayapo people. He was very much one of his tribe's leaders, but he had also travelled extensively to try to let the world know about what was going on in the rainforest. I'll never forget his closing words: 'We are fighting to defend the forest because the forest is what makes us, what makes our hearts beat. Because without the forest we won't be able to breathe and our hearts will stop and we will die.'

It was Paiakan who later suggested to me that we might be able to trade with the Kayapo. The idea was fraught with difficulties, not least the effect that trade with the outside world might have on their culture and way of life, and I was well aware of the danger of disrupting their lives, but at the same time I didn't buy the argument put forward by some environmental groups that the people should be kept in some kind of timeless bubble with no contact with the outside world. That was how they had been left at the mercy of the logging racketeers. Paiakan was convinced that there must be a way of making the rainforest economically viable with a trading strategy that was based on conservation and good husbandry and I was convinced that we could help them. If the controlled extraction of sustainably-managed plant materials could provide a livelihood for the rainforest people, I felt it was up to them how much change they wanted to accept but that they should set the pace and control that change.

All knowledge should be translated into action.

Albert Einstein

We looked at a number of rainforest seeds, roots and plants that might be used for our products and decided in the end that oil from Brazil nuts offered the most potential. I went back to Brazil to meet Paiakan and discuss the project in detail. We flew out to the Kayapo tribal area in a light aircraft. Most of the village people were there to greet us when the aircraft touched down on a little dusty airstrip of packed red earth. As we were walking into the village, a group of young girls came towards us, virtually naked except for strings of red and blue beads worn across their chest like bandoliers. One of them took my backpack and I linked arms with them, laughing with them as they tried to teach me one of their songs. It was a wonderful way to be welcomed into the village.

I was allocated a hut with a hammock near the river and on the morning of my first day the whole village gathered together. Usually the meetinghouse is reserved for men only, but Paiakan managed to get all the women and kids in as well. I had come prepared with samples of raw materials to demonstrate the process of turning natural ingredients into skin and haircare products. I showed them how bananas could be used in shampoo and how oils could be extracted from roots and nuts. Then I asked them to show me what forest plants they used on their own skin and hair.

In the end it seemed that Brazil nuts still offered the greatest potential for our use. When I asked the Kayapo villagers whether they were interested in gathering nuts to make oil for products for The Body Shop, Paiakan assured me that they were. We didn't even begin to discuss how we might pay for them. Cash was not new to the Kayapo. They had sold Brazil nuts before and they were already having money thrown at them by the loggers and they had asked us to come up with a sustainable alternative – which they had never had before. Of course I wanted to be able to reward them as primary producers, but I didn't want to jeopardize their culture. The main thing was that I had to satisfy myself that this could really help them and at the same time stop the logging.

It would be their first sustainable direct trading link with white society and I was proud to be involved in setting it up.

The four days I spent in the Kayapo village were wonderful. I felt more at home with the rainforest tribe than I had ever done in the West. Over the next few years, I would experience that same strange sense of belonging over and over again. Whenever I am with indigenous people, particularly nomadic people, I get the feeling that I've come home. I can't explain it, but there is a sense that things are somehow 'as they should be'.

I bathed with the villagers in the river every morning, spent most of the day learning from them about the forest and slept in a hammock at night. I often went into the forest with the village medicine man, who claimed he could cure any known illness with potions made up from forest plants.

DISILLUSIONMENT

I left the Kayapo happy and excited by the prospect of setting up our first trading deal with indigenous people. Back in Britain, we found a hand-operated machine that could be used to extract oil from the nuts and had it shipped out to them. While we recognized that they would not be able to guarantee a regular supply of raw ingredients, they were great hunters and gatherers. We estimated that harvesting 13,300 lb of dried Brazil nuts every year would produce 3,300 lb of oil for a Brazil nut conditioner, producing a potential income for the community of around $50,000.

In 1990, Gordon and I went back to Brazil to finalize the arrangements. Gordon worked out a simple business plan for the Kayapo and we advanced them the cash they needed to equip the harvesting expeditions and build a drying hut. We were still full of optimism. When I got back to the UK, I wrote:

'What I would ultimately like to be able to do is to set up a perfect example of honest trading with a fragile community and make it a benchmark of how we should conduct such trade in future.'

Our Brazil nut oil conditioner was a great success, but trading with the Kayapo was not easy and it took much longer than we expected. We had hoped that by trading with them for a non-timber product like Brazil nuts we could counter the pressures coming from the illegal loggers who were decimating the rainforest and threatening their homeland. But although the income from the Brazil nut oil business was significant – it went straight to every one of the men and women who harvested the nuts and produced the oil – it did nothing to stop a few of the Kayapo leaders succumbing to the temptation of the really big bucks which came from the illegal loggers. The lure of aeroplane flights, trips to the cities and luxury goods such as satellite television – all provided by the logging companies – proved irresistible. So the logging didn't stop until 1999 in the A-Ukre village.

It took the Indians from A-Ukre almost 10 years to finally turn their backs on the loggers. Many companies would not have persevered, but we did. The Kayapo are proud of their small Brazil nut oil factory, as they call it. They have shown themselves able to manage all the stages of production – from the harvest to the extraction of the oil – without any outside assistance.

Our trade link with the Kayapo survived all the manipulation and false allegations that we were expoiting the Indians. It all started when we discontinued our relationship with a consultant who had been assisting the villages in organizing the Brazil nut oil business. Unbeknown to us, he had been ripping off the Indians for years. When we terminated our contract with him, he started to spread lies. The Indians stood by us and the trade relationship continues to date – and they are proud of it.

After trading with the Kayapo villages for nearly eight years and providing huge amounts of money to keep the relationship going, we were forced to rethink our involvement in the region. With the help of

When you aim for perfection, you discover it's a moving target.

George Fisher

The Body Shop Foundation, all the Indian tribes in the Altamira region – and the Kayapo village of A-Ukre – have set up their own Campealta co-op. This is the trading arm for the villages and its goal is to generate the money needed for better education, independent funding of the health project and patrolling the reserves. The co-op has a volunteer board of advisors formed by professionals from different walks of life who are helping the Indians to get direct contacts with the foreign markets and to devise a business plan. The co-op is allowing the Indians to export Brazil nuts, Brazil nut oil and copaiba oil, a fixer used by the perfume industry. It has also set up a small-scale eco-tourism initiative and an IT provider in the town of Altamira to generate income to the tribes.

Sometimes it takes hard experiences to get the right model. The Campealta co-op is a blueprint for how Indians can gain control over their resources and use them in a sustainable way – and The Body Shop is one of its customers. The other Kayapo villages are now joining in the co-op and for the first time in eight years, the Gorotire Indians have gone out to the forest to harvest Brazil nuts that the co-op will be selling for them.

Campealta is a blueprint of what may become a model that other tribes may decide to follow. Income generated from the different activities goes towards financing their business and also to health and education. The co-op is unfolding as a versatile business initiative. We have learned much in the process about how to manage relationships with communities and I hope this can provide a better business model than the do-nothing, see-nothing predominant model of today.

HEALTH

The health side was more successful. We built one 'health house' in Redenção and in Altamira we constructed and completely renovated an old house the Indians had, which is now the Altamira Indian 'health house'. These were the first hospitals for indigenous people that offered alternative medicine, mostly herbal medicine. We also set up the Green Pharmacy project, a medicinal plant farm and a lab that manufactures over 60 per cent of all the medicines used by the Indians. The Green Pharmacy really works, replacing many of the toxic pharmaceutical drugs with herbal alternatives. It produces everything from natural anti-inflammatory remedies to antibiotics, anti-diarrhoeal syrups, expectorant syrups and much more. Some of the remedies are used by the local hospital as part of a bartering system whereby the hospital provides the Altamira Indian clinic with the pharmaceuticals the Green Pharmacy does not produce. The Green Pharmacy has been incorporated into the Campealta co-op and the plans of the co-op are to increase production and broaden sales to other towns in the region. Demand for natural remedies is growing and the co-op realizes it has a very good business on its hands.

Although the Redenção hospital was forced to close because of lack of support from the Brazilian Department of Health, the Altamira hospital has been an enormous success and serves eight tribal groups, treating some 3,000 patients every year.

The Altamira clinic still assists the Kayapo tribes through the Campealta co-op, providing medicines and other types of social support. This didn't stop the critics from accusing us of exploiting the Indians and only being interested in selling more shampoos.

NEPALI PAPER

One of our first trade links was with 30 producers of handmade paper in Nepal. The Body Shop commissioned a range of handmade paper products from a village in the Kathmandu Valley. Using local sustainable materials, the villagers revived a traditional craft dating back to the eleventh century.

Papermaking in Nepal had declined with the government's restriction on cutting lokta, a daphne shrub and traditionally the main source of paper fibre. Too many lokta shrubs had been stripped and killed, leaving the hillsides vulnerable to the heavy rains which washed away the fertile topsoil. The Body Shop employed a visionary consultant, Mara Amats, to research alternative sources of fibre which at first included water hyacinth and banana fibre but ended up, more prosaically, with cotton rag waste.

Our first range of handmade Nepalese paper was sold through branches of The Body Shop during the Christmas period of 1989. In December 1990, a designer from The Body Shop visited General Paper Industries in Nepal and spent a week working up new designs alongside the producers. We give this kind of technical assistance to our trading partners wherever we can. Not only does this kind of backup produce fine quality products for The Body Shop, but the experience means that the producers can use their own culture and ideas to experiment with new products – not just for The Body Shop but also for other like-minded organizations. Later products reflected the increase in production and design skills since the initiation of the trade link and included colourful photograph albums, frames and scented drawer liners.

The success of the paper products also led to some remarkable local community work. Some of the money that the producers earned was used in their Community Action Fund to invest in healthcare, education and alternative income generation projects. In 1993, 90 young girls were awarded scholarships. Girls were chosen because attitudes in Nepal generally mean that they are not usually seen as a worthwhile investment, so they tend to be married off or – the nightmare alternative – sold into prostitution. As well as educating children, the Community Action Fund sponsored adult classes too.

TRADING SMALL

We put enormous effort into our community trade links. Today, The Body Shop has something like 37 trade links around the world that provide employment for thousands of poor families and I think that's an enormous success. They include sesame seed oil from a village in Nicaragua and cocoa butter from Ghana. The people involved are all getting a fair wage, which means they can reinvest in their own communities. With some trade links, we also pay what we call a 'social premium' – a 10 per cent bonus to help finance whatever community project they choose, usually something to do with education, health or fresh water supplies. This social premium built the first Montessori school in Tamil Nadu, southern India, and I'm still very proud of that.

Community trade demands a lot of time, a lot of devotion, a lot of commitment and a determination to make every relationship succeed, no matter what the difficulties. This is the one area where we are always saying to the competition: 'Hey, come and join us!' Unfortunately, precious few seem to take us up on the invitation.

I suppose there are plenty of businesspeople who believe that no one in their right mind should expend so much energy on something that, frankly, does not have much impact on the traditional bottom line. But these projects do add to our social and environmental bottom lines – and we refuse to measure our success solely by mechanistic profit and loss statements.

I would rather be measured by how I treat the weaker and frailer communities I trade with than by how big my profits are. And if all of us in business committed ourselves to this, big things could really happen. In that sense, if in no other, small is beautiful.

With more than 2,000 stores around the world, we have massive purchasing power – and we want to buy from communities where trade can make a difference. More than that, we want the trade to make a double difference – not just economically, but also through investing in that community. I am a great believer in small-scale economic initiatives. Viewed in isolation these grassroots initiatives are modest – 10 women planting a tree, a dozen youths digging a well, an old man teaching neighbourhood kids to read – but from a global perspective their scale and impact are monumental. These micro-enterprises form a powerful front line in the world-wide battle to end poverty and build a better world.

The US magazine *World Watch* calculates that the world's largest 500 corporations control one quarter of world economic output, but employ less than one twentieth of the world's people. The real backbone of world commerce and global employment is made up of the millions of unsung small enterprises that farm small plots of land, cook food, provide daycare for children, make clay pots, do piecework for apparel makers and carry out countless tasks that larger businesses don't do.

In the cities of developing countries, a growing percentage of the working population is engaged in micro-enterprise activity – sometimes estimated as high as 50 per cent. In Latin America and the Caribbean, more than 50 million micro-enterprises employ more than 150 million workers. In China, the number of small enterprises grew from 1.5 million in 1978 to 19 million in 1991. Nor is the phenomenon of micro-enterprise confined to the developing nations. As many as 90 per cent of UK businesses have fewer than 10 staff. These millions of tiny one-, two- or five-person businesses are going to have to provide a large share of future jobs – and the critical ties to local community and ecology that are so vital to the local economy. They will also provide the main source of independence and empowerment for communities facing the icy winds of globalization. As such, they are what stands between the planet and social and environmental collapse. That's why we are always looking for fair trading initiatives with local communities.

TRADING REAL

Since colonial times, the West's wealth has been determined, I think, by the way we use the majority of the world's labour – our comfort is built on their hard work. I remember passing this huge

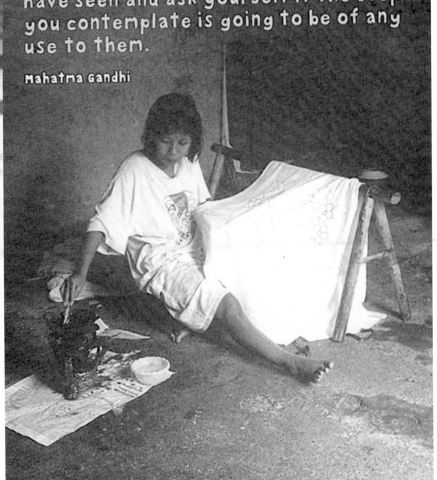

Whenever you are in doubt, apply the following test; Recall the face of the poorest and weakest person you may have seen and ask yourself if the step you contemplate is going to be of any use to them.

Mahatma Gandhi

field in Bangladesh where they were firing bricks. The men were so thin that their bodies were as wide as my arm, but they were running with hundreds of bricks on their heads from one side of the field to the other, as if they were in a marathon.

Now there are so many basic things we no longer know how to do because our society has become so automated and mechanized. Where is the human factor in all that we do? What fascinated me was seeing villagers in Third World countries actually making things with their hands. They weren't reliant on machines to make their products and in fact there was little or no electricity, so even energy had to be generated with the power of their hands and their feet.

I can't help feeling that if there were more of these products, if more companies invested time and effort in trading with communities in need, we might eventually begin to combat poverty. In this way, communities work together, their skills are protected and families can stay together. A human dimension is brought into the work environment. And the products are of superb quality. So you get the whole resonating package, rather than an anonymous purchase that comes with an invoice.

When we buy a product, be it an ornamental pot or a decorative basket, do we ever question its origins? Few of us have any idea of how a pot is fired without electricity and this lack of knowledge underlines our complacency as consumers.

We need to ask how, why, where and when things are made. Knowing more about the generation and source of products empowers us.

TRADING FAIR

Consumers are increasingly aware that their purchases are moral choices as well – they are actually responding faster than most businesses. They crave knowledge; they yearn for information that supports

business as unusual

radical choices, which in part accounts for some of the success of businesses like cafédirect and other fair trade organizations. Consumers want the stories behind the products. Why, as our Western economies flourish, are these stories so often miserable and empty? Why must they be stories of sweatshops in Los Angeles and Asia, growing ranks of child labourers, rivers full of effluent or political prisoners jailed for questioning the local effects of corporate power?

The Body Shop believes that trading should be an ethical act. Fair trade is absolutely central to us. It means we have to avoid the direct exploitation of humans and animals and avoid any negative impact on their habitats. And knowing that environments, and the people in them, are not exploited in getting products to market means we can give consumers the information to choose more responsibly.

To help us determine whether a potential supplier could benefit from a trading relationship with us, we set up some criteria to help us make sure we are trading fairly with communities. Our fair trade guidelines say that small communities we trade with must be:

☞ socially or economically marginalized
☞ involved with and benefit from the trade
☞ commercially viable
☞ able to build a trading relationship that can benefit the primary producer or processor
☞ using a product or process that is both socially and economically benign and sustainable

The first criterion is, of course, that the community wants to trade with us and has something we can incorporate into our product range. I have already described my strange experience with the Native American tribe, the Oglala, who invited me into a sweat lodge and eventually decided not to use sage bushes in haircare products. I was more successful in Ghana when I was given a fabulous opportunity by the BBC to film a programme about grassroots women's enterprises in Africa. Naturally, I jumped at the chance and during the course of the filming I met the resourceful women's association that now supplies us with shea nut butter.

GHANAIAN SHEA BUTTER

I've never looked on Africa as a poor continent. To me it is tremendously rich in culture, in traditional skills and in ways of caring for the body unknown in the West. The women in Tamale, in the north of Ghana, have been using shea nut butter as a skincare product and for cooking for generations. Shea nut trees grow in the north of Ghana, where the ground is dry and the vegetation is sparse. The women collect the nuts from the trees locally, then pound and grind them into butter using traditional skills that have been handed down over the centuries. It is a long and arduous process, but they know instinctively when the butter is ready.

I wanted the women to make the butter for us because I knew that if the nuts were simply collected and exported, the women would not get the value-added price they could command by processing the nuts themselves. The greater the price they could command, the greater the reinvestment in the community. Further down the road in this trading relationship, that social premium contributed towards paying for a school and much-needed medical services.

Initially the women offered to 'fair trade' whatever butter they produced in excess of their needs, but they were soon able to acquire rudimentary machines that enabled them to produce much larger quantities than the traditional methods allowed. That said, there is still a lot of manual work to be done using their ancient skills, although the actual pounding and grinding of the nuts can be done by machine. I feel a little ashamed that we rarely, if ever, make that connection between what we buy and the truly backbreaking processes that go into making it.

The women in Ghana do the vast majority of the work in their communities and they are exactly the kind of women I instantly fall in love with and want to spend hours talking to. What I also love about the spirit of these women is that they haven't allowed themselves to become dependent on technology. What do they do when the machine that kneads the butter in a minute needs a replacement part that has been on order for months? Like most women, they just improvise and get on and do it by hand – a process that takes about three hours.

I went back to visit these women not long ago. They come from 10 different villages so we did a lot of driving in 35° heat across dusty, dry terrain to meet them all. They came out to greet our car, singing and dancing down the road towards us. In every village I was given gifts – yams, guinea fowl, goats – and because these gifts were an honour, I couldn't refuse them. So off we'd drive in the heat for another two or three hours with a couple of goats on the back seat. Goats stink – but then I've always said there's nothing like knowing the story behind your product.

INDIAN WOOD AND COTTON

One of my favourite trading relationships is called simply Teddy Exports, after the son of a remarkable woman, Amanda Murphy, who used to be a member of staff at The Body Shop in Oxford Street, London. Amanda fell in love with an Indian and moved to India to set up Teddy Exports. The name speaks volumes about the business, which is like a close-knit family – run from the heart.

Teddy Exports runs a factory in Tirumangalam, in Tamil Nadu, which makes wooden and cotton products to sell in our shops world-wide – footsie rollers, twin ball massage rollers, cotton bags with quotes on. It employs 265 men and women directly and another 75 indirectly. All are fitted with free uniforms, lunch, tea and healthcare for themselves and their dependants. They also have daycare facilities, free transport and a Montessori-style school and, thanks to the little wooden footsie roller, they've managed to build a new wing for the school. The project has also bought a house in the middle of the town where women are taught sewing skills so they can get jobs and family planning is available for the larger community.

Gordon and I once spent a week around Valentine's Day in Tirumangalam, which the locals say means 'good fortune'. Sounds romantic? No chance. Years ago, Tirumangalam provided most of the roses for the holiest city in southern India, Madurai, but as the drought has got worse, the roses have been replaced by thousands upon thousands of trucks stopping to refuel. There is plenty of sex to be had, but little in the way of romance and there is a huge problem with AIDS.

For the larger community Teddy Exports provides education in family planning, sexually transmitted diseases and AIDS awareness education through the Teddy Trust, set up from the profits of Teddy Exports. Children are welcomed into the classrooms of non-formal education centres. And even animals are catered for, too, with veterinary clinics. Along with Soapworks in Easterhouse, Teddy Exports is one of the community connections I am most proud of in all my years at The Body Shop.

HEMP: THE WONDER CROP

Of all the thousands of ingredients that can be found in The Body Shop products, I don't think there is one that is more controversial or more potentially beneficial than industrial grade hemp. In the same way that we pioneered the use of jojoba, camomile and *Aloe vera*, The Body Shop is the first major retailer to incorporate the benefits of hemp seed oil into skincare products. It's not likely to be the last.

In 30 years in the cosmetics industry, I can think of few ingredients that stand out simply because of their performance and efficacy, but hemp is one of them.

The politics of conscious-
ness.

Description by The Body Shop of its campaigning activities

Hemp is the most versatile, eco-friendly, economically viable crop in the world, but it suffers from a bad case of mistaken identity, being frequently confused with its cousin, marijuana. Suggesting that industrial grade hemp products will send us down the road to reefer madness is about as sensible as saying that Poppy Day promotes opium abuse. Industrial grade hemp and marijuana are both members of the cannabis family, but they are botanically distinct. Ganja, weed, pot, marijuana, whatever you call it, is a narcotic drug. Its psychoactive properties are derived from the chemical Delta 9 tetrahydracannabinol (THC). **The leaves and flowers of the marijuana plant contain up to 20 per cent THC. Industrial grade hemp, on the other hand, contains less than 1 per cent. You'd need to smoke a joint the size of a telegraph pole to get high.** More than that, hemp actually contains a relatively high percentage of a cannabinol, CBD, that blocks the marijuana high, so it is actually an 'anti-marijuana'. And it hits the weed where it breeds. If marijuana cross-pollinates with hemp, it loses its potency. So there is no incentive for marijuana growers to conceal their crop amidst hemp fields, which is one of the reasons cited by officialdom for hindering the cultivation of industrial grade hemp.

I first became aware of hemp when we were looking to develop a new range of lotions for dry skin. A member of the staff in the Phoenix branch of The Body Shop and my daughter in Canada both asked me at around the same time what I was doing about hemp. The answer was nothing, because I knew nothing about it, but the more I learned the more I realized what a wonderful ingredient it was. Hemp seed oil has an unusually high concentration of essential fatty acids, which makes it an ideal lotion for people with very dry skin, but it has about 25,000 other potential uses – everything from building to brewing beer to fuel. The trouble is that almost everyone thinks it is also a mind-bending drug.

The negative attitude towards hemp is a relatively recent phenomenon – history has been much kinder to the plant. The oldest existing example of human craft is thought to be a piece of hemp cloth dug up in what was once Mesopotamia and dating back to 8000 BC. The sails of Columbus' ships were made from hemp cloth. Rembrandt painted on hemp canvas – in fact the word 'canvas' is derived from 'cannabis' – and so did Van Gogh some years later. America's founding fathers, George Washington and Thomas Jefferson, were hemp farmers. The first pair of Levi's was woven from hemp fibre. In 1929, the auto industry pioneer Henry Ford was investigating the possibility of building and fuelling a car entirely from hemp – I could go on and on.

The fact is that, by February 1938, the hotbed of radical thought *Popular Mechanics* had identified 25,000 possible uses for hemp. So what went wrong?

Public perception of hemp had been influenced by sensationalist articles linking crime and insanity with marijuana ever since 1915, when several New England states had passed anti-cannabis laws. Social reformers of the time feared anything which might release inhibitions. Ironically, the federal prohibition against alcohol, which came into force a few years later, meant that the market for cannabis increased. The negative propaganda continued, however, with William Randolph Hearst's papers associating 'marijuana madness' with 'degenerate' Mexicans, African Americans and jazz musicians. Then the world went to pot in the 1960s and suddenly the word 'cannabis' rang one big alarm bell in the corridors of power.

In 1961, the UN Convention on Narcotic Drugs recognized the difference between industrial grade hemp and marijuana. So did its 1990 supplement. But in the meantime, America's Comprehensive Drug Abuse Prevention and Control Act, passed in 1970, had effectively made all cannabis cultivation illegal in the US, with a zero-tol-

THE EXPRESS

SATURDAY MARCH 7 1998   4?

FLY TO
EUROPE
FROM

£38

RETURN

● FIRST TOKEN PAGE 77
Subject to Express allocation

Virgin Express

PLUS: Play INSTANT WIN £100,000 Virgin Mega Hot Tickets

0171-629 4817

BODY SHOP IN DRUGS STORM

EXCLUSIVE: Roddick accused of cashing in on 'cannabis chic' over cosmetics made from hemp

BY JOHN INGHAM

HEMP
Cannabis

BODY Shop chief Anita Roddick caused uproar yesterday over plans to sell skin creams made from the cannabis plant hemp.

She was warned that the move could legitimise the drug in the eyes of young and impressionable customers.

Invitations to the launch of a new range show a picture of the distinctive cannabis plant with the slogan: "Hemp, it's a growing phenomenon. Will it grow on you?" Individual cosmetics also carry the cannabis symbol.

Multi-millionaire Ms Roddick, Britain's most successful businesswoman, is a leading supporter of a campaign to decriminalise the drug.

Last night one senior politician accused the Body Shop of acting

"irresponsibly". The products, which include soap and lip conditioner, are made from industrial hemp, a member of the cannabis family.

Hemp production is banned in several countries, including the United States and Australia. In Britain, it is only grown under licence from the Home Office.

The Body Shop stressed that industrial hemp contains only a tiny percentage of the "mind-

TURN TO PAGE 2, COLUMN 2

UNDER ATTACK: Anita Roddick and, left, the hemp symbol shown on the new Body Shop range

216   business as unusual

erance policy for THC at any level. But America's words and deeds simply haven't matched up. As recently as 1994, an Executive Order from the President designated hemp 'a strategic crop of importance to national security', though you certainly wouldn't know it from the US drug czar General Barry McCafferty's attitude. Hemp is 'a novelty product [sustaining] a novelty market', he said in 1997.

Some novelty! Consider industrial grade hemp's environmental advantages. Because hemp is naturally resistant to bacteria and pests, it can be grown without insecticides and herbicides. So there is no toxic run-off from hemp fields. By way of contrast, the world's current top fibre crop – cotton – uses nearly 30 per cent of the world's pesticides. Industrial grade hemp produces several times more fibre per acre than a typical forest and matures in just 120 days, so widespread use of hemp would reduce deforestation. The annual world consumption of paper has risen from 14 million tons in 1913 to over 250 million tons in the '90s, which means that about half the world's forests has been lost since 1937.

Hemp's longer fibres permit hemp paper to be recycled more extensively than traditional paper and it can be pulped using far fewer chemicals. Hemp suits most climates and it grows fast, yielding two harvests a year. It only takes 30 days for the plant to put down a root system of 10–12 inches, which breaks up soil compacted by overworking. In the UK, farmers who planted hemp as a rotation crop found they got a 10 per cent increase in yield the following year with wheat. Hemp also grows so densely that it smothers weeds and reduces pests in the soil. As it matures, its leaves fall and this encourages soil fertility. And the root system retains 70 per cent of its nutrients, so it makes good fertilizer when it is ploughed back under.

Two things recommended hemp to us at The Body Shop. First we could create an innovative product range and secondly we could re-educate our customers about hemp's many virtues.

Real creativity, the kind that is responsible for breakthrough changes in our society, always violates the rules. Richard Farson

business as unusual

Hemp fibres are excellent for textiles. One company has produced a line of hemp and hemp-velvet textiles for such makers of furniture and apparel as Ralph Lauren and Tommy Hilfiger. Other companies are combining hemp fibres and lime to make lightweight cement and plaster. Hemp seed itself contains 30 per cent more oil by volume. Stable at a wide range of temperatures, hemp seed oil has been used to make high-grade diesel fuel oil and aircraft engine and precision machine oil. Hemp seeds themselves contain all the essential amino acids and fatty acids necessary to maintain life. No other single plant source provides complete protein in such easily digestible form. Hemp seeds can be crushed for oil and then processed into flour for cakes and bread. Hemp oil can be used to make cheese, milk and ice cream.

So we had a remarkable ingredient with a long history of human use – and a bad image problem. We had no idea what a furore we would cause.

BIZARRE REACTIONS

We created a range of hemp soap, skin oils, creams and lip balm and designed the packaging to look industrial. We wanted to launch the range at the Chelsea Physic Garden, but we were refused permission to plant any seeds because the garden is not allowed to grow hemp. In the end we launched the product with a bunch of synthetic hemp plants, but immediately ran into trouble.

At the event to launch the range in Britain I handed out the seeds and was accused by former Home Office minister Ann Widdecombe of turning drug-taking into a joke. Ms Widdecombe said I was 'wholly irresponsible' because the hemp products derived from the *Cannabis sativa* plant. Parents Against Drugs also protested.

'Do you honestly believe the sight of a hemp plant will drive Britain's youth to drugs?' I asked when I was questioned about it by the press.

In Canada, where the industrial grade hemp had been grown for us, we had to glue the products to the shelves during the launch to prevent the press from smearing the products on themselves, hoping to get high from using too much. In the United States, ludicrous restrictions were imposed on the launch and the word 'hemp' had to be obliterated on all the packaging. In Hong Kong, the launch ran foul of the territory's strict anti-drug laws and we were forced to withdraw the whole range after tests by government scientists claimed the products contained traces of THC.

In France, gendarmes descended on The Body Shop in Aix-en-Provence and 'arrested' the hemp lip conditioner, hemp hand protector, hemp multi-purpose oil and hemp elbow grease, along with promotional material, claiming it would 'incite the use of a banned substance'. When I was asked for my reaction I said: 'I know the French perfected the art of irony in the past, but right now I'd like to see them get a better grip on the future.' The range was temporarily banned in France, but suspicions were later assuaged.

I suppose, in the cold light of day, you have to be grateful for the crass stupidity of these reactions. It did give us some bizarre amusement as well. But at the time, I started to play the conspiracy theorist.

It didn't make rational sense to me that the extraordinary benefits this plant can supply were being so vilified. It is only ignorance that confuses hemp with marijuana.

One botanist said that trying to extract THC from industrial grade hemp was as difficult as extracting a trace element of gold from the sea.

But hemp's potential enthrals me, which is why I am so infuriated by the confusion that surrounds it. Things aren't as bad in Europe as they are in America or Australia. At least it is possible to grow the stuff in Europe under licence. But the hemp industry in the UK is threatened by a proposed 25 per cent cut in EU subsidies to hemp farmers, and producers say no other crop is being targeted in this way.

In the end, I expect it will be money that talks loud enough to guarantee the hemp story has a happy ending. The most optimistic projections for hemp products predict annual sales of $5 to $10 billion in the early decades of the twenty-first century. Can you imagine any politician with his own best interests at heart opting to miss out on a slice of that pie?

And in the United States, the sustainable farming of industrial grade hemp would be invaluable in halting the disappearance of the small family farm.

We hope our new hemp range will not only benefit our customers, but will also help an old and valuable industry get back on its feet.

WIN SOME, LOSE SOME

I can't pretend that The Body Shop trading practices offer a solution to everybody's needs. But in committing ourselves to community trade, we believe we have created a way of trading that will satisfy not just the needs of our customers, but of our business as a whole – and also our trading partners in marginalized communities. I know we have a lot of work to do to improve understanding about these trading relationships among our customers, and even in some of our own shops. But the concept of community trade is fundamental to The Body Shop credo. It may take time, but I believe we are creating a model of consciousness, a new paradigm of how to treat the economically weak and frail.

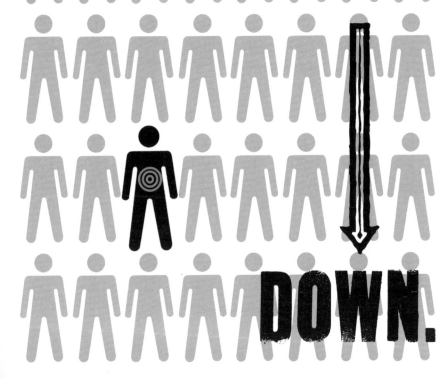

THE DOWNSIDE OF BEING CONSIDERED

SOCIALLY RESPONSIBLE

IN ANY AREA IS THAT THE PUBLIC AND THE MEDIA TEND TO PLACE YOU ON A PEDESTAL FROM WHERE YOU HAVE NOWHERE TO GO BUT...

DOWN.

A BULL'S-EYE ON MY BACK

The day that Channel 4 chose to broadcast a documentary that savaged The Body Shop and effectively accused Gordon and me of being cheats, hypocrites and liars, I was struggling home from one of the most difficult and dangerous trips I have ever undertaken.

It was May 1992 and I had been in Sarawak. My daughter Sam, then living in Vancouver and allegedly studying homoeopathy, had alerted me to the plight of the Penan people, who were fighting the loggers destroying their rainforest. They were organizing human blockades to hold back the loggers and Sam and I decided we'd try and get some film of their protests to alert the world to what was going on and, as usual, help subsidize them. We booked under assumed names with a travel group that was a covert indigenous activists' group and flew over rainforest that stretched from horizon to horizon until it gave way to red earth flanked by rivers running red. This was our welcome to the land of timber concessions and selective logging.

To connect with the Penan, we had to trek for 14 hours across difficult terrain. To say I was unprepared was an understatement. Plimsolls are for boat decks, tennis courts and city streets – anywhere but primary jungle. But I doubt that anything could have prepared me for the leeches that balanced on the edge of every leaf,

Alan Redert, *Best 75 Business Practices for Socially Responsible Business* ready to jump onto any exposed skin. Or the cobwebby roots that tripped me up every few seconds. Or the immense sense of claustrophobia, shut in by jungle and hardly able to see the sky. Or crossing ravines and rivers on logs. It was an utterly exhausting journey. Hours went by without a word from anyone but Sam, who seemed oblivious to the hardships. She chattered on to the others about *The Jungle Book*, sang songs from movies, talked about Gandhi, the ozone layer and vigilante consumerism. By the time we reached the Penan, she had a blister as big as her foot. I couldn't stand, sit, raise a leg or hop. I popped a sleeping pill and fell asleep to thunderous rain, dreaming – not altogether happily – of crossing ravines on logs.

The next day we sat in on a meeting as Penan leaders discussed the possibility of another blockade. Their biggest problem was food, because during the blockades they had to leave their farms. I gave them some money so the village could repay its debts and we got some good film. The journey out was just as bad, through torrential rain. I was worried all the time that we could be murdered by the loggers. They didn't find us, but the leeches did – over and over again.

On the flight back to Britain, covered in leech bites, I was too tired to think about anything but getting home. I had been warned that the press were waiting for me at the airport, but I had no idea

what it was about. British Airways managed to smuggle me off the plane to avoid the waiting reporters and Gordon was there to meet me. He looked drained. He told me the Dispatches programme on The Body Shop had been broadcast the previous night. 'You're never going to believe this,' he said. 'They're saying that The Body Shop is nothing but a sham.'

It was early in 1992 when representatives from a company called Fulcrum Productions approached us about a programme they wanted to make for Channel 4. They said they wanted to do a fly-on-the-wall documentary about a socially responsible 'feel-good' business and assured us they would produce a fair and balanced film. We agonized for a bit and finally decided it would be an excellent idea. We gave them absolute co-operation, total freedom. We let them film at our headquarters in Littlehampton and at Soapworks. Both Gordon and I were interviewed and we gave them volumes of printed material, in-house videos, films and stills.

They had full access to the company. But what they produced was a shameful piece of television reporting that defamed both the company and us. Maybe it wasn't their motive from the start, but television is about tension and I think they realized that if they could find something – anything really – that would make The Body Shop look hypocritical, they would have a brilliant programme they could sell in every English-speaking country in the world where we were high profile. I believe it was a business decision on their part.

We were also vulnerable because we were so high profile. We were openly challenging the system, challenging the role of business. We were campaigning, we were loud-mouthed, we had attitude and we were making waves as well as big profits.

Reputation, reputation, reputation – the one immortal part of man. *Othello*

It was obvious, when I look back on it, that they were looking for evidence that showed us up as charlatans and hypocrites.

AN ATTACK ON OUR REPUTATION

I was devastated when Gordon showed me the video. It is really eerie to see something or to read something about yourself and realize that you don't recognize yourself at all. The programme was a vituperative attack on the company's core values and everything we stood for. It suggested that our campaigning was no more than a cynical marketing ploy to deceive the public and generate sales and attacked the personal integrity of both Gordon and me.

I couldn't believe we could be seen that way. There are no circumstances in which I would compromise my reputation about animal testing for profits. Our policy was set out in leaflets in every shop explaining that animal testing was a complex issue and that the best we could do was abide by the rule devised by the British Union for the Abolition of Vivisection (BUAV): that no ingredients had been tested on animals within at least the previous five years, by ingredient suppliers or anyone else. But the programme alleged we were really no better than others in the industry who thought little of testing their products or ingredients on animals.

The programme makers also claimed that our community trade initiatives were suspect. We were more used to criticism from that direction – the left was hitting us all the time with these kind of accusations, claiming we were destroying the culture of the indigenous groups, and the right hated us because we were progressive in our thinking. Even so, *Dispatches* was the first and the biggest attempt to challenge the reputation of The Body Shop.

I believe you live your life on your reputation and this was an attack on the one thing that had we built our entire business on. After a little while, the duplicity of the programme simply made me angry.

But then there was another sense in which we were exactly right to be targeted. We were so comfortable with what we thought we were doing, or what we perceived we were doing. We needed a jolt. We just didn't need a fully-fledged frontal attack based on falsehoods and half-truths.

We felt we were doing good things, in an open way, so what harm could we possibly be doing?

It struck me as so ironic: here we were trying to do things decently and honestly and being accused of the exact opposite. The trouble is that unethical businesses aren't as big a news story as perceived fallen angels and hypocrisy. And I think some of the press had always wanted to witness the decline and fall of The Body Shop because I don't think they fully understood what we had tried to do. We certainly had our enemies in the business and financial press.

OUR RESPONSE

On the day after the programme was aired our shares dropped from 270p to 160p. There was also a dramatic downturn in sales for a while because many of our customers – at least those who believed the programme – felt they had been betrayed. But it wasn't the financial damage to the company that worried me so much as the loss of our reputation and the damage to the faith that people had in the company.

Gordon and I decided immediately that we had no option but to sue. Everything we stood for was on the line. Gordon knows the law really well and would never normally have considered embarking on something as risky as a libel action, but in this instance we had to. The press has to tell the truth and we had to defend our integrity. There was no way we could have done anything else. **For a while we were hot news. This was a really big story, because our entire reputation was at stake.** We set up a 24-hour communications operation so that our staff around the world – Australia, New Zealand,

I would rather go to any extreme than suffer anything that is unworthy of my reputation, or of that of my crown.

Queen Elizabeth I

Canada, anywhere – could monitor and challenge misleading press coverage. We had the press camped outside our offices in Littlehampton and outside my home. At one stage there was such a crush of media outside our offices that it was difficult for the staff to get in and out. I remember one day there was a BBC truck with a huge radio tower hanging microphones out over the roof of what they thought was the boardroom. It was incredible.

While this was going on, we were trying to repair some of the damage, but the timing was wrong. Some groups like Greenpeace and Friends of the Earth came to our support. But we were also at the point where a lot of the alternative trade associations thought we were meddling buggers. They believed we shouldn't have gone to the indigenous peoples and set up direct trading links with them because that was their job. Because we had been *ad hoc* and spontaneous, we got less support from them.

GOING TO COURT

The trial was at the High Court in London and it was a nightmare. I walked to the court every day. I had so much pent up anger and energy that I couldn't take a cab. I have never walked so fast in my life and the angrier I got, the faster I walked.

Being in a courtroom for six weeks was like being incarcerated in a mahogany coffin with alternating periods of incredible tedium, tension and nervousness. Thousands of documents, internal memos, position papers and videos were brought before the jury and Gordon and I were in the witness box for days. In the witness box I fiddled unconsciously with the buttons to my blouse, and I heard later that friends watching – not to mention our lawyers – were willing me not to undo the next button down and unexpectedly reveal myself.

There were a couple of memorable moments. At the end of one day of testimony, the head of our Against Animal Testing department Rita Godfrey got exasperated with the opposing counsel. She said: 'Look, if you think I would spend eight years of my life working in Against Animal Testing just as a PR job, you must think I'm crazy.'

Against that, of course, was all the legal language that is so full of conditional clauses and double negatives that it is sometimes hard to extricate yourself. You have to say, 'Hang on a minute, am I saying "yes" or "no" here?' I remember at one point, counsel asked me something and I just said: 'I can't remember.'

I never believed we would lose the case. I could not believe that if you are honest and you tell what you perceive to be the truth and you say it with passion, that the truth wouldn't prevail.

He said: 'What do you mean, you can't remember?'

I said: 'Look, you've got five bloody barristers around you, 65 solicitors, all telling you what to say, don't tell me about remembering.'

The judge ticked me off and said: 'Keep to the point, keep to the point!'

But after a 27-day trial we were completely vindicated. We were awarded £276,000 in lost profits and £1,000 for each libel and Channel 4 was served with an injunction forbidding them to repeat any of the defamatory statements. This was extremely important to us because we knew that they had been planning to distribute the film around the world. All in all, with costs awarded against them, the case cost Channel 4 about £2 million.

VALUES AND VISION DEPARTMENT

We felt we had won the trial on our reputation and our values, so we had to find a way of incorporating this into the fabric of The Body Shop. In the months of preparing for the trial, our lawyers had access to everything. They reviewed every internal memo to determine what we were doing to 'walk the talk'. This process showed us that there wasn't complete solidarity in the

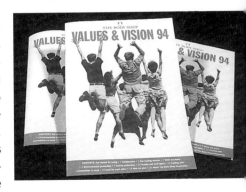

ranks. Best practices were not being followed uniformly throughout the company. We realized we needed to take stock and touch base with our staff around the world.

After we returned from London, we set up a huge meeting in the warehouse which every member of staff from our headquarters in Littlehampton attended. We told them why we had won and what our values are and how we were going to capitalize emotionally on the victory. Shortly afterwards we set up the Values and Vision department to integrate the commercial and ethical dimensions of the company. We made community trade more effective, wheeled in a human rights group and added the campaigning department. The aim was to shape the company into a really socially responsible entity. We needed to institutionalize it because I was frightened that all this *ad hoc* stuff we did was going to be seen as just an add-on.

The most damaging thing *Dispatches* did to us was to make us timid and put the legal department in a position of unwarranted power.

After *Dispatches* we felt we had to ethically audit everything. If you wanted to fart, you had to have it audited.

THE CORPORATE STALKER

Then, four weeks after the end of trial, the man I would come to know as the Corporate Stalker showed up in Littlehampton.

The Corporate Stalker is an American journalist who became obsessed with The Body Shop, in particular with the idea that we were running an entirely disreputable business behind a veneer of social responsibility. He clearly believed he could make his name, and perhaps his fortune, by exposing us. He seemed to be a constant presence in my life for years, following me around the world, showing up at meetings where I was due to speak and doing his best to discredit us. He was short, bearded and sweaty. People whose opinions I respect have told me not to give him any more publicity by naming him here, so I won't. But his presence in our lives meant that the trial victory against *Dispatches* wasn't quite the end of having to defend ourselves against public attack on our values. It was just that the Corporate Stalker took the whole issue to such ludicrous extremes that it would have been funny if it hadn't been so dangerous to the company.

ABC'S *PRIME TIME* DIDN'T FAULT US

When he first approached us he said he was working for ABC's *Prime Time Live* and he had been commissioned to do a piece on The Body Shop. Despite the *Dispatches* débâcle, we still believed that our best policy was to co-operate fully with journalists – after all, we had nothing to hide. So we gave him hundreds of pages of documents and offered to provide more or less anything he wanted.

But within a few days we began to get worrying feedback from the suppliers and franchisees he was contacting. It was soon pretty obvious he was working to an agenda and was looking for material to discredit The Body Shop. His technique was to harass people until he got something he could use. He called one of our suppliers more than seven times in two days. Whenever he got hold of somebody he would misrepresent what another person had said in an attempt to elicit the responses he wanted, and by doing so managed to create a maelstrom of misinformation and fear.

He told people that The Body Shop was engaged in a 'covert policy of testing on animals'. He told them he had 'documented evidence' that The Body Shop was suffering from massive problems, that it was a 'sick company' and its franchisees in North America were 'in revolt'. He told them that our Trade Not Aid policy was bogus and that our 'manipulation of indigenous people had provoked hatred and mistrust', that we had violated environment laws and that our Harlem community project was 'worse than Easterhouse'. This last allegation was strange since our Soapworks factory in Easterhouse was a big success and was distributing 25 per cent of its cumulative post-tax profits to the local community. **I suppose some might have seen it as ironic that we were eventually forced to use what corporate influence we had to fight back.** So we contacted ABC to tell them how their reporter was behaving. The next thing we heard was that he had left ABC with the programme unfinished. He would add his departure to his list of grievances against us, claiming we had 'intimidated' the network. It is hard to believe a major American news organization with a reputation for hard-nosed investigative journalism would be intimidated by the likes of us.

THE FDA DIDN'T FAULT US

But our troubles with the Stalker were far from over. While he was still at ABC, in September 1993, he had gone to the US Food and Drugs Administration (FDA) with a number of allegations, leading inspectors to make a surprise visit to our headquarters in Wake Forest, North Carolina. No citations or violation notices

What a weak barrier is truth when it stands in the way of an hypothesis. Mary Wollestonecraft were issued. Undeterred, he was still giving the FDA the impression he was working on a programme about The Body Shop for ABC, and he got them to make a second visit. Again, there were no citations. Then, using the Freedom of Information Act, he got copies of internal FDA notes of both visits and leaked them to other journalists as 'proof' we were being investigated.

There was one problem with this ploy: all the FDA internal documents made it clear their inspections were initiated 'in response to allegations made by a *Prime Time* news reporter'. He simply blotted this out in the documents he leaked.

VANITY FAIR DIDN'T FAULT US

After he left ABC, the Stalker took his story to *Vanity Fair* and started telling people it was going to be a cover story. As soon as we heard about it, we protested to the magazine for assigning an article to a man clearly working to an agenda. Again, I make no apologies for doing so, or doing anything we have done to protect ourselves from someone hell-bent on doing us as much damage as he could.

We were pleased when we heard that *Vanity Fair* had dropped the idea. It was clear to us by then that, for reasons that we could not begin to fathom, the Stalker was engaged in a single-minded campaign of vilification against us. He contacted investors and brokers, telling them what he was about to publish would cause our stock price to fall, which was bound to destroy investor confidence and drive down our share price. He turned up at a meeting in Toronto of the Social Investment Forum and harangued the assembly about The Body Shop until he was told to sit down. He then handed out copies of the grossly defamatory article which he had written for *Vanity Fair* and which had been rejected by them. He circulated the same article to social activists, fair trade organizations, animal rights advocates and others in Europe and the US. He called Jay Harris, the publisher of *Mother Jones* magazine, and described us as 'the most evil corporation [he had] encountered in 20 years of reporting'. He was quoted in another newspaper saying his investigation would be 'the story of the century' – putting us right up there with world wars, the Holocaust, Vietnam, the end of the Soviet Union and countless human and ecological disasters.

He prepared and circulated 'dossiers' on The Body Shop to NGOs, fair trade organizations, animal welfare campaigners and many others. He turned up at business and franchise meetings to denounce us. He harassed me in public and in private, even pestering me at home. To anyone who would listen he described The Body Shop as 'morally despicable', 'vicious on the inside' and 'the most hated company' he had ever come across. At one point, the police were called to remove him from one of our shops in the United States after he started shouting at customers. He contacted a number of the witnesses involved in the *Dispatches* trial, accused them of perjury and suggested that there would be a retrial.

FIGHTING THE ALLEGATIONS

After all that effort, the Stalker's campaign began to have some impact. In August 1994, it was reported that the US ethical investment fund Franklin Research and Development was selling 50,000 of The Body Shop shares in the light of 'accusations' about our ethical policies – mainly our commitment to Trade Not Aid, the use of oil-based ingredients in our products and emissions from our former factory in New Jersey.

All these allegations, curiously enough, were the substance of the Stalker's rejected *Vanity Fair* article and grossly misrepresented the truth. Franklin was selling because the fund managers thought we were facing increasing competition and that our share price was too high. Reference to 'emissions' from our factory almost made it look as if we were dumping nuclear waste, but in fact what had happened was that we had accidentally spilled some 65 gallons of shampoo, washed it down the drain and rung the authorities to tell them so. It was a pity, but it was hardly the *Exxon Valdez*.

Even so, the story was picked up by the *Financial Times* in London and started a veritable firestorm in the British media, bolstered by leaks from the Stalker, which generated widespread anger and frustration throughout the company. The Body Shop was the subject of nearly 150 articles in under a fortnight, almost equal to Bosnia or Rwanda. As we were subjected to trial by headlines, we worked round the clock, exhausted, see-sawing between exaltation and disgust, straining to exhibit grace under fire, startled by the feverish personal tone of much of the coverage. I tell you, reading about yourself as others see you on a daily basis – even when they're trying to say good things – is disorienting, even fiercely frustrating. But it's not nearly as frustrating as countering a war of Chinese whispers.

It was like shadow boxing, or trying to nail water to the wall. Everything was alleged, nothing was directly stated.

Repetition was the key. Repetition gave weight to the lies and rumours. Add to that the popular conviction that there's no smoke without fire and you get more than enough fertilizer for wayward media conjectures. I picked up the Guardian one Saturday to read that the RSPCA was advising customers to boycott The Body Shop because our Against Animal Testing policy was dubious. Never mind my belief that policy is ultimately rendered irrelevant by practice, the RSPCA 'recommendation' was a figment of someone's imagination. Still, it was 'reliably' reported along with the rest of the innuendoes and we were left gasping at our lack of recourse. The RSPCA itself subsequently corrected the story in a letter to the newspaper.

Anything that bolstered the idea that we were hypocrites was served up. I was accused of 'tainting' the company by agreeing to appear in three high-profile American Express advertisements. The charge was that I had abandoned our 'never advertise' stance and given way to the pressures of the competitive US market. No one thought to ask me about the ads or find out that I had only agreed to do them after American Express had promised me that I could talk about community trade. I subsequently donated my fees to indigenous people.

THE LESSONS I LEARNED

Carl Jensen, the American friend who runs Project Censored, which annually publishes the 20 most censored stories in America, told me he'd never seen the media jump with such alacrity on a story with so little proof. 'There is no doubt in my mind that if you'd spent as much money on advertising as you do on social causes, this deluge of criticism would not have occurred,' he said. It was what we got for not playing by the rules, for endeavouring to change the way business could be done. **But I probably shouldn't have been surprised. Any company that is values-led can expect extreme responses – either elevated above the angels or cast down with the demons, with a bull's-eye glued to your back to boot.** It is a fact of life for anyone – individuals or organizations – who has stood up for what they believed in. And God help you if you are a woman.

That was the first lesson from the whole affair. Another was that 18 years of hard-earned reputation counted for absolutely zero against the perceived newsworthiness of a concerted attack on our integrity – mine and the company's. It was almost as though the press believed their case could only be convincing if it was purely negative and presented in a mock-grudging 'We are not happy saying this but we feel we must in the interests of the truth' manner that would do justice to Shakespeare's Iago.

I was taken aback by the glee with which critics personalized their attacks on the company by focusing them on me. I couldn't help but think that a male executive wouldn't have been subjected to quite so intimate an assault. In one newspaper I was portrayed as a Mother Teresa figure. It struck me as distinctly odd that I should be measured against her standards, rather than against chairmen of companies responsible for seriously heinous corporate crimes. Another newspaper suggested I must be kicking myself for ridiculing traditional City figures as 'pin-striped dinosaurs' – I actually also called them 'a load of wankers' – because I needed these same dinosaurs to weather the crisis. That wasn't true either. **But then passion does breed passion, as I have always said. The trouble is that truth wasn't served very well in this case.**

BUSINESS ETHICS

The Stalker's voluminous research, on which others based their allegations, was dubious. He concentrated on finding flaws and then he tried to shape what he'd found – which was often from far in the past – into a sinister web of present-day activities. By the simple law of averages, I would have thought it was likely he would have uncovered strengths and successes as well, but there was no acknowledgement of them.

His approach to business ethics seemed to demean the whole idea. It rested on the assumption, shared by so much of the business press at the time, that 'business ethics' was actually an oxymoron. Therefore, if any attempt to raise ethical standards was impossible, by claiming an ethical core we must have been hypocritical – or so the argument went.

We were pretty sure that no responsible magazine would publish the Stalker's so-called 'story' and so we were particularly concerned when we learned that *Business Ethics*, a small-circulation publication which is the leading media supporter of responsible and ethical business, was planning to publish a version of the article that had been rejected by *Vanity Fair*. **I simply couldn't understand it. I was bitterly disappointed that a publication which claimed to be searching for new business**

paradigms should show itself to be so profoundly rooted in old business values.

We pleaded with them not to publish, explaining that we had been fending off the unsubstantiated allegations for more than a year. When *Business Ethics* refused, we asked for permission to submit a response for publication in the same issue. They said 'no' to that request too. They even refused us an advance copy so we could prepare a response to the media.

Business Ethics published 'Shattered Image' in September 1994. Editor Marjorie Kelly said she was publishing with mixed emotions: 'We have been ardent admirers of Anita Roddick and her company for many years; two years ago this month we featured her on our cover. But after weeks of debate, we concluded the greater good would be served by raising these issues in print.' We would have been more impressed by this sanctimonious nonsense if we hadn't known that before publication Ms Kelly had sent a letter to investors in the magazine saying the piece was 'the best thing we have ever done' and 'could put us on the map'.

The article was so obviously a hatchet job that in reality it turned out to be something of a damp squib, particularly after all the hysterical build-up. An analyst at Goldman Sachs said the allegations were neither new nor serious. Others said it didn't add up, that The Body Shop was still 'considerably kinder to animals, the environment and people than many other listed companies'. And many of our friends rallied round, including Ben Cohen, co-founder of Ben & Jerry's, who resigned from the magazine's advisory board in protest. 'This unbalanced, questionable piece of journalism does not advance constructive dialogue about social responsibility,' he said.

Jonathon Porritt was similarly supportive. 'Ever since Anita became a leading figure in the green movement, people have been gunning for her in one way or another,' he said. 'It's an absolute classic syndrome in this country that success of any kind, be it in the environmental world or elsewhere, means that some people spend most of their lives trying to knock you off your perch.'

Ralph Nader also weighed in, calling the Stalker 'the mouth that roared' with 'an inclination towards fiction'.

'Maybe the next time,' Ralph suggested, 'he should get into the trenches and go after the likes of GE and Monsanto.'

DAMNED IF WE DO, DAMNED IF WE DON'T

My personal reaction to the crisis was to eat like a pig. Anything and everything. Chocolate for breakfast, nibbles all day, lots of water. Nothing you needed a fork for – I just had to keep stuffing things in my mouth. I also found myself swearing a lot. But in my message to the staff I put on a bold front:

You can either take or leave the challenge to run a company in ways that make a difference. We take the challenge. All we are trying to do is bring our heart into the workplace. It needs adaptability, courage and a love of change. It also needs the dedication of a convert. Make no mistake. We have entered uncharted territory: there are no signposts out there, there's no map. No one's been there before. But if we don't keep trying, we are no more than bystanders.

When the dust at last began to settle some observers accused us of over-reacting. *Time* magazine claimed that 'each side outdid each other in shrillness and paranoia, obscuring any reasonable point either side had to make'. We were still quite sensitive in those days and we probably did over-react, but we wanted to make sure that if there was going to be a backlash against what we were doing, it was at least based on fact rather than fantasy. As I told one journalist not long afterwards: 'Look, a gun was pointed at my baby's head. I'm not going to say, "Fuck you, do what you like." I'm going to behave like a lioness with her cub.'

But, as Gordon said at the time, we're damned if we do and damned if we don't.

The occasional rap on the knuckles is good for us. It makes us look at what we're doing to see how we can improve it – but I did object to someone trying to demolish everything we'd created in the last 18 years.

Everyone at The Body Shop knows that life is an experiment – we all just aim to go in the right direction if we possibly can and to do our best. Running a business like The Body Shop was a constant compromise between idealism and practicality – but I still don't believe we deserved to be attacked in the way we were. Funnily enough, that same year happened to mark the moment the deeply conservative Institute of Directors finally came out and said things I had been saying for years: that business had to move with the times and introduce feminine values like compassion, intuition, creativity and caring into the workplace. And at the company's AGM that year, a motion that The Body Shop should enshrine a commitment to human rights and social responsibility into its articles of association was carried unanimously.

I eventually got hold of a copy of the Stalker's original unpublished article and it was very odd. It described me as an 'anti-capitalist icon with a rambling Georgian mansion, a flat in London, a castle in Scotland and a personal fleet of cars'. At the time I probably earned less than any CEO of a comparable company anywhere in the world. But my 'fabulous rags to riches' success story hid a 'far more sinister tale of broken trust', according to the Stalker. We were 'knee deep' in fraud investigation by the Federal Trade Commission and the Food and Drugs Administration, our animal testing policy was a 'sham' and our Trade Not Aid projects were 'deceptive and colonialist'. Doing business with The Body Shop was, the Stalker said, like 'dealing with the Gambino crime family'. And, strangest of all – given the competition out there – he dubbed us 'the most evil company in the world'.

Enough said.

THIS IS DRIVING ME NUTS.

THE LEGAL DEPARTMENT SAYS
WE CAN'T USE THE WORD
'ACTIVIST'
IN THE FOUNDATION
ANNUAL REPORT BECAUSE
OF ITS ASSOCIATION WITH
TERRORISM!
THIS REALLY IS TOO
PYTHONESQUE FOR ME –
**I'M GOING TO USE
IT WHEREVER I CAN.**
IN FACT, I'LL NAME A FRAGRANCE
AFTER IT.

Activist

THE LEGO SET FROM HELL

I was giving a talk in London to a group of management personnel and one of the women in the audience asked me a question that stopped me in my tracks. 'Why do you keep calling The Body Shop *your* company?'

It was a damn good question. My first reaction was that I did so because it *was* my company. I was really possessive about it. It was me that went to bed and worried passionately about it. Companies are like children: you give birth to them, they grow up, get married, have a life of their own, but they're still yours and you would no more hand them over to strangers than you would your real children or grandchildren. But equally, nobody had ever asked me the question before. The Body Shop may have been mine in the sense that I started it, but I could no longer make all the decisions and as it grew larger and began to change the only tangible power I had over it was the force of moral persuasion.

And if it wasn't mine exactly, whose was it? The employees who had spent two decades shaping it with me? The franchisees who had made such an enormous contribution to its success and values. Both of those certainly, but with them the same thing applied – there was no way I wanted to hand them over to strangers or outsiders either.

My diary in 1996, before the Lego set from hell process

Even so, as The Body Shop grew in the 1990s it became clear there were organizational changes we badly needed to make. And if there is one mistake I regret more than any other during the roller-coaster ride of my career, it's allowing myself to be persuaded that outsiders could come into the company and tell us how to run it. I should have known from the outset that it was stupid. I wouldn't expect an outsider to tell me how to run my life, however expert they were. Yet it was one of these consultants who organized and reorganized The Body Shop into such a complex structure that I began referring to it as 'the Lego set from hell'.

Running a company is rather like a marriage. When it's going well, it's fantastic. When it's not, it's absolutely miserable. Around the time of the *Dispatches* programme and the hornet's nest stirred up by the Corporate Stalker, it wasn't going at all well. The media barrage which we had to put up with during this period led to a lot of corporate introspection and soul searching about the kind of organization we were. This in turn resulted in a form of internal McCarthyism as our rapidly-growing legal department vetted everything we did. It was at this point that they ruled out the word 'activist' in the annual report of The Body Shop Foundation because of its association with terrorism.

The creativity, camaraderie and sense of fun that had marked our early years were seeping away. It used to be that we had no one-, five- or ten-year organizational plans. But by now I had accepted that the massive evolution of the company meant we needed a change in the very fabric of its leadership and management.

GETTING A MANAGER

We appointed a managing director to set up policies and systems, increase profitability, make operating decisions and generally take over the management of the company from me. I think we felt it was the required 'spoonful of bad medicine' to get the company organized and functioning smoothly.

When the new managing director first took over I was relieved. I was grateful I didn't have to attend to the minutiae of the business any more. But there were immediate problems about control. His first act of professional management was to make the first redundancies in the company's history. It was only 25 people, but it fomented enormous distrust and fear. I was away when it happened. I had wanted to communicate the news of the redundancies myself because I knew what a devastating impact it would have, but he just went ahead and announced them. I was furious.

The new MD also set up five different profit centres and created competition within the company, but it seemed to me that he only succeeded in driving a wedge into the sense of family which was so cherished by staff. The company was being organized around geography and territory, which I thought was very male thinking. **It was threatening to turn into the kind of company I said I would never be a part of.**

Littlehampton was increasingly the focus, rather than the shops and the staff – and certainly not the customers or the community. And at the same time as this reorganization was taking place, the recession began to bite and our annual profits dropped by 15 per cent.

THE PROBLEM OF POWER IS HOW TO GET MEN OF POWER TO LIVE *FOR* THE PUBLIC RATHER THAN *OFF* THE PUBLIC

ROBERT F. KENNEDY

It wasn't long before I took against our new managing director. He felt he was installing efficiency. I felt this 'efficiency' was leading to an intense system of patronage and power. I felt his methods frightened the staff. The Littlehampton headquarters was soon rife with in-fighting and competition and the unique qualities of The Body Shop were becoming steadily neutralized. We had our first face-to-face confrontation at the time of the Gulf War when he tried to put a stop to my 'Stop the War' campaign (see Chapter 4). I won that one, but at the time I wondered what the future would hold for me if we were constantly at each other's throats.

But I needn't have worried. While I was out of the country, he called a board meeting to discuss what was to be done about the 'wild card' in the company's organization – by which he meant me. Fortunately for me, he seriously underestimated my support. The board felt it was my style and ideas about trading and retailing that drove the company forward and gave it a future and they felt very protective of me. After his challenge to my position, the MD had little alternative but to leave himself.

It was at this point in our development that we were blinded by the light of the $2 billion management consultant industry which was out there beckoning to us. 'Come this way,' the light was saying, 'and you too will be fixed.' So, we came, we saw, we spent – handsomely as it turned out – and we put our faith in the words of strangers. We were psychoanalysed and deconstructed and reconstructed and put into little airless boxes. We lost touch with our heritage, with who we were.

We pulled something apart in order to meet a challenge we didn't understand. It was a big mistake.

Ichak Adizes was an Israeli-American based in Santa Barbara, California. He was a highly-respected pioneer in management behaviour and consultancy who had achieved considerable prominence by devising a 'lifecycle' theory for companies. Adizes' theory is that there are appropriate and inappropriate problems at each

Most of us are pathetically grateful for anything that is run properly.

Alistair Mant

stage of a company's development. The companies that deal with the appropriate problems move on to the next stage of growth; those that don't eventually die.

We first heard about Adizes when he was featured in an article in *Inc.* magazine in the US. One of our board members was very enthusiastic about him and said we had to meet him, although the process of deciding to hire him was actually a very loose affair. I think I had a quick conversation with him in San Francisco, but in situations like that people tell you what you want to hear – things like: 'This company has got to be preserved. We've got to make sure it stays, we've got to get your values through, we have got to get it institutionalized, we have got to make it all work.' My initial reaction was that I was for anyone who could get us out of the mess we were in, but I was frightened that change would result in losing the soul of the company.

Gordon was much more positive about Adizes than I was. Gordon thought his approach to corporate change and restructuring was appropriate for us and that the sheer size of our business meant that we needed a forward-looking plan.

Gordon said: 'Nobody is looking ahead for three years. We are not looking at the possibilities. If we don't, we won't be here in three to five years.' And he was probably right.

The relationship between Gordon and me had bequeathed a very distinct management style to the company – loosely structured, collaborative, imaginative and improvisatory – and this matured as the company expanded. I think Gordon provided a sense of continuity and constancy while I bounced around breaking the rules, pushing back the boundaries of possibility and shooting off my mouth. We rarely argued and when we did it was never about values. But as the company grew there was a growing feeling that the sense of family was being diminished. Quite apart from anything else, the bigger we got, the less direct contact every member of staff could have with Gordon and me.

But then businesspeople have an obsession with 'plans' in a way that isn't shared by most of us. We don't make five-year plans for a marriage or five-year plans when we have children – and when we do, who keeps to them or even remembers them? People like Stalin used to organize fearsome five-year plans, but they were entirely for the benefit of the outside world and the results were faked. Complex systems need planning, of course, but it's absolutely crucial that you don't lose your flexibility. You have to be able to respond to the unexpected and the very nature of plans means that they can't take account of that. **I love tactics; I don't like plans.**

I didn't worry about the decision to take on Adizes straightaway, but serious doubts set in after about a month. I was afraid he wasn't reflective about what the company needed and that he didn't believe in dialogue or communication. He seemed to just barge ahead with reforms, working to a formula that might well have been successful with very different types of companies, but which I wasn't sure was appropriate for us. He seemed to me to see the problem through a traditional and hierarchical model of how companies used to be run. He seemed to assume that solving a problem meant going straight to the top. His style may have worked brilliantly for an organized, manufacturing company, but it doesn't work nearly so well when the value of a company is based on creativity.

THE DREADED ORGANIZATIONAL CHART

Adizes wanted to restructure the business with endless organizational charts. Under pressure from me, the main chart at least had our Values and Visions department at its centre, so that the 40 people dedicated to championing our values inside and outside the company could make sure they were enshrined in the new structure. The main problem was that I am usually the kind of person who is constantly trying to break down structures. The notion of hierarchies is very uncomfortable for an entrepreneur.

As the pre-eminent destroyer of inert structures, it was probably inevitable that Adizes would bring a great deal of tension with him, particularly between Gordon and me. I told Gordon I just didn't like him, couldn't see his value for us and thought he was wrongheaded for a company like ours. For Gordon it was a matter of principle – we should give him a chance. **But then suddenly six months had gone, and God knows how much money, and we were in a bigger mess than when we started.**

Adizes' theory was that the company was metamorphosizing from what he called the 'Go-go' phase to 'Adolescence'. As he put it:

We now have reached a stage where the idea is working, the company has overcome negative cashflow and sales are up. It appears that the company is not only surviving, it is even flourishing. This makes the founder and the organization arrogant, and arrogant with a capital A. [I assumed this was a reference to me.] Now in adolescence, the organization is being reborn apart from its founder – an emotional birth. In many ways, the company is like a teenager trying to establish independence from family. This rebirth is more painful and prolonged that the physical birth of infancy. The most distinctive characteristic behaviour of adolescent organizations is conflict and inconsistency.

This is how we explained his theories to the staff:

The Body Shop is now 18 years old and according to Ichak's definitions of a company that means we're still an adolescent – and we all know what that means ... you fall in and out of love like there's no tomorrow, you're passionate, vocal, anarchic, full of energy and yet you're still having to come to terms with physical and emotional changes over which you feel you have no control. What is important for us at The Body Shop is to correctly channel all that energy so that we operate at our peak – Ichak calls it 'Getting to Prime'. His training is putting a strategy and discipline into reinventing how we get on with the business, without losing touch with everything that makes us different, and without losing our spirituality. He's helping us look *inwards* at the way we all work *together* which is so fundamental to our success in the future – not just here in the UK but world-wide. The Body Shop community is well formed – what we've still got to do is push the grand experiment forward and reinvent ourselves.

But the staff were worried about Adizes's influence over our core values. He was an unknown quantity, a mysterious 'messiah' brought in from California, and was greeted with suspicion, although his track record was impeccable – he had orchestrated the evolution of dozens of companies with as many as 90,000 staff. But when I looked back on the whole episode much later I realized that nowhere in our conversations with him was there ever any mention of values or of what was sacrosanct to the company.

Gordon did his best to create a link between Adizes' organizational changes and our values while I struggled to make sure that he understood our commitment – I even sat in excruciatingly dull meetings on economics. But the differences between us were soon obvious to everyone. Very early on, Adizes was in the middle of a mini-lecture about his methodology, explaining that it would allow The Body Shop to grow comfortably into a £1 billion business, when I interrupted him.

I don't give a toss about being a bigger company,' I said. 'I care about becoming a *better* company, a more values-driven company.'

He didn't seem to understand what I was talking about: he didn't seem to recognize the tension I saw between growth and 'losing your soul'. **For me, he seemed to be threatening all the things I held dear. For him, the things I held dear were an impediment to progress.** We did try and discuss the differences between us, but that only meant he didn't talk to me for a lot of the time.

I had hoped Adizes would institutionalize the values, institutionalize the creative process, safeguard the idiosyncratic and experimental thinking, offer a way of getting more transparency in the company and capitalize on the notion that ideas from the staff could provide the answer to many of the problems. I was hoping that all the things I held dear would be held sacred within the company but at the same time it would be run more efficiently, with more speed and less distraction from its purpose.

Adizes' first initiative was to conduct a series of meetings, with a cross-section of the staff and franchisees, where they had to write down all the bad things about the company and come up with ways to correct them. He would collect reams and reams of information about problems and then slot them into sections under a range of different headings. His whole approach seemed to me to be largely about the negatives. I felt it would have been more constructive to be positive, to invite the staff to identify what made the company special, what was worth saving and what was worth striving for. Adizes was very good at macho organization, macho exploits, but he was less impressive in dealing with creativity and women.

I thought what we needed was a human facilitator, rather than somebody who wa capable of organizing the entire Israe army.

I felt Adizes' approach was not honouring the company. His language was alien to us. He was secretive, loud, ruthless and table-thumping. He gave out information without any sensitivity or compassion. I was also embarrassed by the way it seemed the board was buying into his methods, as if he were the one true answer to all our problems. I couldn't believe how the company just took to his ideas like a dog to his bowl of water. And I was wondering what had happened to our native spirit and sense of anarchy.

Most of all, I hated the way Adizes dealt with the employees. I can remember him stamping his foot, scratching his crotch, eating an apple or a sandwich, while exclaiming: 'You! Go out of the room while we discuss you!' His singling out of people, his confronting them, his reorganizing, his brutal shifting of people were respon-

we're against
testing cosmetics
on animals.

{ using despots, however, is another story. }

⟨⟩ THE BODY SHOP

FOR A FREE CATALOGUE, CALL 1-800-BODYSHOP OR VISIT WWW.THE-BODY-SHOP.COM

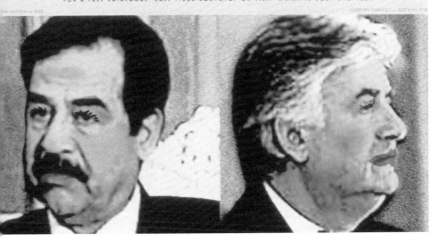

sible for the worst case of corporate anguish we had ever experienced. He wanted to move people out very fast – some of them people we should have protected, people who were part of our culture and had helped shape our corporate identity. To my absolute shame, I didn't protect them. I will go to my grave wishing I had and wondering why I failed to. But the argument was always that I interfered too much, and I still never know – even now – whether to interfere or just shut up and let other people run the company.

I remember a meeting in a hotel when Adizes called a lot of people together and brought them into the room one by one and just said to them: 'I don't think you should be in the company.' **Gordon and I watched this process in a state of shock. I think that was the point when I truly began to understand what we had bought into. I didn't like it at all.**

We should have cut and run then, but we didn't. We carried on with these endless 'Potential Improvement Point' meetings. We were supposed to call them PIPs, and we'd gather these together, then try to work out a strategy for incorporating them. So there was a lot of gathering and a lot of talking and then there was supposed to be an action plan for implementing them. But where was it?

It was a very unhappy time. Gordon and I seemed to be warring all the time and I was battling with the board. There was a disconcerted sense within the family of the company, as if 'Mum and Dad' were fighting. People were becoming disillusioned. Sales were slowing down and there was a feeling that the company had lost its way, that it wasn't being as creative as it ought to be. I became obsessed with trying to stop Adizes and recruiting support against him.

I think disillusionment with Adizes manifested itself among the staff as disillusionment with the board. It was hardly any surprise that at our first Values meeting after the *Dispatches* case it became clear the staff felt the company's strong sense of family was being threatened. This was their conclusion:

There exists a lack of faith in management and lack of feeling of personal empowerment. The staff does not know how much they can say, or to whom; this

business as unusual

leads to lack of risk-taking and a reluctance to be accountable. Staff are unable to speak out because they are expected to 'be positive and toe the company line'.

After 18 months of agonizing soul-searching, Adizes' tedious process of reorganization finally resulted in our first semblance of an organizational chart – the notorious 'Lego set from hell'. There was only one problem: it didn't work. All he seemed to have succeeded in doing was eroding our values and our language. When the board met to discuss whether or not we should renew his contract, there was really no discussion. We all knew we had made a mistake trying to fit a business with a distinct social agenda into the straitjacket of the standard disciplines of management, marketing, finance and operations.

ENSHRINING THE VALUES

It was painful, but I also learned a great deal from the process. I learned that there is an enormous demand from employees for information, care-taking, compassion and dialogue which must be honoured – because at times like this you don't move with one heartbeat. I also learned that we had to enshrine our shared values of honesty, respect and care for people, animals and the environment, and that if we didn't they would become no more than a hollow add-on and we would be no different from any dime-a-dozen cosmetics company. Our employees feel that the values of The Body Shop are part of the company's DNA and if you mess around with the values, you mess around

But putting our future into the hands of a management consultant was the worst mistake we had ever made, and it was our fault. We had made the wrong choice.

with the company's reason for existing. I also learned that, as we grew bigger, it was tremendously important not to lose our sense of humour. We had to learn to live with and laugh over our mistakes.

We should have spent more time searching for a management consultant who was right for our kind of company – someone who would say: 'My God, you have got an image, an edge and a reputation that we have got to preserve.' But then what we really needed was more creativity and fewer consultancies. The whole process had cost us about $2 million.

After Adizes left, the structure he had created was dismantled and a new one put in place. The discussions following his departure produced our mission statement, which was designed in part to prevent any other outsider coming into the company and trying to overturn our core values. We didn't call in an outside PR agency to produce the mission statement, we got our staff around the world to write it – in a process which must have been like the flurry of letters which produced the American Declaration of Independence. I remember being in Australia and watching the different drafts pouring out over the fax and feeling proud that we were writing something with more nobility to it than the narrow bottom line.

The mission statement calls for The Body Shop to:

☞ dedicate our business to the pursuit of social and environmental change
☞ creatively balance the financial and human needs of our stakeholders: employees, customers, franchisees, suppliers and shareholders
☞ courageously ensure that our business is ecologically sustainable: meeting the needs of the present without compromising the future
☞ meaningfully contribute to local, national and international communities in which we trade, by adopting a code of conduct which ensures care, honesty, fairness and respect
☞ passionately campaign for the protection of the environment, human and civil rights, and against animal testing within the cosmetics and toiletries industry
☞ tirelessly work to narrow the gap between principle and practice, whilst making fun, passion and social care part of our daily lives.

We finally got back to our roots through our self-esteem campaign (*see* Chapter 8), which was met with an immediate recognition in our customers as well as in sections of the media. It seemed we had once again done something right at the right time. The feedback was phenomenal. Customers wrote in to tell us they were ours for life. From depressing introspection we had begun again to look outwards – and with style.

But even without the Lego set from hell, we still had the original problems: we had grown complex and our sales were beginning to fall away. We were shocked to find we had lost around a million customers in about three years.

The organization might have been complex, but the problems were simple.

THE PROBLEMS

THE COMPETITION PROBLEM

Everyone and his dog were producing The Body Shop-style products. Even Richard Branson had launched Virgin Vie and in the high streets Boots, Tesco, Marks & Spencer and Sainsbury's had all shamelessly latched onto our format and were selling similar products in almost identical packaging. The big cosmetic companies were the same. Estée Lauder had launched a new beauty range, Origins, which came in recyclable containers, had a high natural content and was not tested on animals – where can they have got that idea? Aveda was producing cosmetics made from distilled plant and flower essences and also boasting of its commitment to the community, with employees being given time off to work with local schools.

There were also a lot of new companies entering the market. When we started out we invented a new concept – a shop in which you could buy anything for the bath and the body and the hair. Nobody had done that before. The department stores couldn't really compete with us because they had 20,000 other products to sell. In fact, very few

Business is like riding a bicycle. Either you keep moving or you fall down.

John David Wright

people tried to copy our concept until we went to America. Then, because of the rise of the fashion industry and strong brands like Jigsaw and Gap, everybody thought the easiest thing they could do to shore up their brands was to make toiletries and perfumes.

Then you had the rise of the hairdressing culture, with the accompanying hairdressing products. If you go into a hairdresser's nowadays, half of the shop is retailing their own products. They had a customer and made her look good and so it was easy to sell her the product too. This wasn't happening 15 years ago. But now things were being shaved away in unexpected areas.

People would say: 'Oh, your competition is Boots.' Boots wasn't the competition: *everything* was the competition – hairdressers, department stores, the fashion industry, the perfume industry. Even bookshops like Barnes and Noble introduced their own fragrances. I know they say that imitation is the highest form of flattery, but this flattery was hurting us.

THE HURRY PROBLEM

We were suffering from a bad case of 'hurry sickness'. Our *modus operandi* was to open shops. **It was: 'Let's open a shop! Let's have one here! My friend wants to open up a shop! Why not?' And it was like that all round the world.**

We just thought it was so easy to open shops, because people were just pouring in. There was no time to pause and analyse, to pull out something and examine it and put it back. We had no time. Nor do I really understand why we were hurrying. It certainly wasn't to please the stock market, as one critic suggested.

I used to ask why we couldn't act like the Romans did when they invaded England? They advanced and they consolidated, advanced and consolidated. I think that strategy would have been better for us, but we were always under this pressure to perform. And we never seemed to have that much respect for our own knowledge – we were always thinking other people knew a better answer.

THE CREATIVITY PROBLEM

We thought of ourselves as a shop-opening company, rather than a marketing or product-led company, so we hadn't changed the look of the shops or the style of the packaging for 10 years. Retailing is fashion and people may have been a little bored with ours. We had created one of the world's most recognized brands – The Body Shop was listed twenty-seventh in the world – without putting a penny into advertising, so we were doing something right, but it meant nothing without the sales to back it up.

Nor were we always supplying our millions of customers with the products they wanted. What they wanted from us was imaginative, idiosyncratic products rather than generics or another peach bubble bath. We should have been putting money into guerrilla creative thinking, with our creative antennae listening to what was going on in the community. In the community where my daughter Sam was living in Vancouver they had been screaming for years about alternative medicines, self-help and home education. There they don't use plastic cards as money, they barter and exchange. They're the ones who had been telling us about hemp for years. If we had had an organized system of listening to those groups, we would have brought in hemp years earlier. We wanted to churn out millions of bottles, to use the filling capacity in the manufacturing area as much as we could, and we didn't spend enough time on the creative process. We had these vast machines and they cost a lot of money, so we wanted to keep them working in order to maximize their potential. But it was making us less and less flexible and less able to adapt to the needs of the market.

What we needed was an avalanche of ideas that kept us separate from the competition. Instead we got absorbed in manufacturing.

All of a sudden we found our retail offering was dictated by our need to recover the overheads of the plant. And because the manufacturing side was related to the retail operation, it never felt the absolute requirement to deliver anything on time or to cost. Also, because it had only one customer – us – we really didn't have the option of using product ideas from other manufacturers or reducing costs by going elsewhere. I was traipsing round the world looking for new ingredients in the souks and markets and finding great stories. Then I'd come home and say: 'Hey look! You've got all this stuff that I've collected around the world, but you aren't doing anything with it.'

We needed to be less scientific about our marketing and products, more challenging and more imaginative. We needed to be a more opportunistic company, much more light-footed. We had become an operational giant, too big to move.

MAKING THE CHANGES

The moment when we recognized that we would have to change more radically came about seven years ago, when a combination of flagging sales and currency problems in the Far East made us face the fact that The Body Shop was no longer a unique, new, exciting and challenging entry into the market. Except for our campaigns, we were perceived as boring and mainstream. The shops and product lines were looking the same and we weren't reinventing ourselves strongly enough. And at the same time we were employing more and more people. If one person didn't have the skill to do the job we wanted done, we would simply get someone else to do it as well – sometimes there were four people doing the same job.

FINDING A MANAGER

The first problem we faced was finding a new chief executive. We realized we would have to look outside the company because we had grown too fast to nurture sufficient expertise and to some extent we had never really focused our minds on who might take over from Gordon and me. We knew it wasn't going to be easy, because we're a complicated, emotional company. We didn't want a traditional

What is required at one stage is the equal and opposite of what may be required at the next stage in the company's growth.

Geoffrey Moore

businessman from a company like Shell: our values are more like those of a church or a temple than a business. We had also had 23 years of a founder-entrepreneur in charge, which was a huge worry to many people who might well have been interested. In fact, 90 per cent of the people who were headhunted expressed concern about my role and suggested that if they were to come in and do the job well, Gordon and I should have significantly reduced responsibilities. **I didn't understand why I should be classified as 'part of the problem', as the headhunters put it.** Neither Gordon nor I are in the least interested in the mechanics or disciplines of business, but we're both skilled at certain things. He is a great networker and a great venture capitalist. He can see an idea faster than anybody and he has the commitment to put money into it and bring in the expertise. I'm interested in style and image, communications and social responsibility. None of these areas of expertise would mean we would be interfering in the running of the company. I also thought it was odd that the people we headhunted imagined I did not have good commercial sense.

Despite the headhunting problems, we knew the kind of CEO we wanted to recruit. We didn't want a retailer because we thought such a person would be too narrow in approach. We wanted someone with broad-based international business experience – someone who had seen and done a lot, and someone who had enough balls to stand up to Gordon and me.

It was a strange process I found myself essentially handing over the company's well-being to a stranger. I have to say, looking back, huge mistakes were being made. There wasn't enough reflection time and not enough due diligence done. When asked that were the biggest mistakes we made, other than going public, it was the process of replacing ourselves.

TELLING THE MEDIA

When we finally appointed a new CEO and the announcement was made, it unleashed a completely unexpected media assault on me, claiming that he had 'replaced' me and that I was being forced to 'step down because of pressure from disappointed shareholders'. It was ridiculous – I wasn't the CEO anyway – but in the next few weeks it was open season and all the problems of The Body Shop were laid at my door.

eople were stopping me in the street to say how sorry they were. I'd go to the chiropodist or hairdresser and they'd say: 'Well, how does it feel to be retired?'

It was clearly another lesson in humility, but at the time I was curious as to why we had allowed the financial press to believe that I had been replaced. My job hadn't actually changed at all – I was still doing what I had always done.

REORGANIZING THE COMPANY

Some time before our new CEO arrived we had decided that the company itself needed a major reorganization so that we could concentrate on product development, marketing and retailing. The company was too cumbersome and top heavy.

It was never a mistake to go into manufacturing: it was probably the only way we could have supported our incredible growth. We had started manufacturing in a small way about five years after the inception of the company. Our original idea was simply to protect the integrity of our products – to make certain, for example, that none of the ingredients had been tested on animals. But as the company expanded, we were manufacturing more and shipping much greater distances. The distribution costs grew and the products started getting too expensive by the time they arrived in the shops. We had three layers of margins – we wholesaled to the head franchisees and then the head franchisees wholesaled to the sub-franchisees. By the time the customer got the product, it was about a third more expensive than it should have been. We also seemed to be totally unable to get a new product out into the market in less than six months. Most retailers were getting new things out in six weeks.

When we announced we were getting out of manufacturing, we had a lot of people coming to Littlehampton wanting to buy the plant. What we wanted to do was find somebody who would keep the site and keep the jobs – because we were the second biggest employer in the town – and hopefully create more jobs. In the end we made a deal with a great group from South Africa, all guys in their early thirties who are dynamic and used to contract pharmaceutical manufacture and who also espouse our values.

Of course, we then had to face the problem of enforcing ethical standards on second- and third-party manufacturers, but we put a strategy in place for that. We called it 'due diligence'. To protect the integrity of the supply chain, the company won't start relationships with new manufacturers unless they undertake a full social, animal and environmental audit. Suppliers are asked to complete a screening questionnaire which tells us all we need to know from the environment, health and safety and quality points of view. That's the way outsourcing works for the company now.

We still create the product ideas in our own laboratories, or working with other laboratories, and then, we find someone to manufacture them for us, to our precise specifications and wherever in the world is most suitable.

My next step to change the way business is carried out was to combine business and education, which is sadly cut off from the real issues of life, like poverty, ethics and creativity – with a sense of moral purpose and creativity. In 1994 I helped found the New Academy of Business together with the Centre for Action Research in Professional Practice at Bath University. The New Academy is trying to bridge the gap between visionary business leadership and what is being taught in conventional management schools. There is a real need for business education to champion values-aware management. Business schools need to move away from the emphasis on short-term stockholder-based solutions, away from the concept that business is either about profit or principles but not both.

The idea came about after I was invited to be guest lecturer to the MBA students at Stanford University and it was a profound lesson for me. Here was all this huge intelligence, but no guidance about new business thinking, nothing on human rights or corporate codes of conduct. The only explanation of ethics the students seemed to be given was an instruction not to take bribes. I felt business education needed something better.

The New Academy is now working with businesses, universities, NGOs, UN agencies and other partners, both nationally and internationally, to continue to research, explain and articulate a different model for business and to help create a just and enterprising future. **Our vision is to help build the next generation of business education based on responsibility, accountability and respect for diversity. It's a revolutionary step.**

I THINK AS LEADERS, WE SHOULD BE MAKING SURE THAT BUSINESSES SHIFT THEIR EMPHASIS ON THINGS TO AN EMPHASIS ON THE HUMAN SPIRIT.

I DON'T BELIEVE

THE MOST VISIBLE CHANGES IN COMPANIES OF THE FUTURE WILL NOT BE IN THE SO-CALLED 'SCIENCE' OF BUSINESS, BUT IN WHO WORKS FOR YOU, WHY THEY'RE DOING IT AND WHAT THEIR WORK MEANS TO THEM ...

REINVENTING
THE BODY SHOP

Reinvent for survival – that was our mantra in the late '90s. Reorganize, rework, re-engineer – whatever the word was, we had to change. We had to change our profit stream from manufacturing and distribution to sales and marketing, set up four distribution, manufacturing and marketing units across the globe, do all this seamlessly and still be able to supply all our stores.

When you reorganize a company, it is usually inevitable that some people's jobs won't be there at the end of it. And so it was that in January 1999 we were forced to announce that the reorganization would result in job losses both on the manufacturing side of the business and in management. It was a time of trauma and anguish for the whole company. It seemed that the 'golden age' had passed and we were behaving in a more traditional corporate way.

You can tell a lot about a company by how it tackles the tricky and miserable business of making people redundant. My constant question to the board was why redundancies first and not as the final option. The answer was that we'd over employed. What I learnt during this time of corporate anguish was if we didn't enshrine our values into the entire company they would be seen by everyone whom we employed or supplied as hollow. Now, I look back at that time with immense pride – pride at the thoughtfulness and kindness that went into the redundancy process. It was a world-class act.

At first, we didn't know how many people we had to make redundant and we didn't know how many people were keen to get into the new structure. It surprised us how many of our people were willing to take voluntary redundancy – but they were getting a phenomenally good package. Many of them were also single and under 30 and just wanted to go travelling or set up their own enterprises.

The next stage was that those who had to be made redundant were helped by their managers and the human resources department to work out what their next working life plan would be. They were offered training in a new skill which they could apply somewhere else and training was also offered to any other member of their family, if they chose. Sometimes a female employee said she wanted to give up work altogether, but told us that her husband could use the training.

Then we formed an entrepreneur's club to help the people who wanted to set up their own enterprise – whether it was a consultancy or a café – with advice and interest-free loans. Some of those we helped start up have gone on to amazing success stories of their own.

We also worked with all the social service and business development agencies in Littlehampton and we had continuous open discussions between town and staff. We talked to the churches, community workers, social workers and the town councillors so that they knew what was happening from the beginning. Everyone was so well prepared that I don't think any of the counselling services were ever needed or called upon. And everyone who left had full use of all our facilities – from the childcare centre to the gym – for another nine months.

I don't think we could have handled the situation any more thoughtfully or more diplomatically than we did, but I think, on reflection, we definitely missed out on compassion, on spending more intimate time with people, especially the temps, some of whom had been with us for over a decade. It was a terrible period and it caused a great deal of heartbreak. On a video we made to explain what was going on to the employees I looked shell-shocked, and that was how I felt. We weren't like big industries who make people redundant all the time

– we were a family and offered wonderful jobs with wonderful conditions. To accept that some of those jobs had to go was a big wrench – especially as we had built a 23-year reputation for creating jobs rather than destroying them. We lost about 300 people in the end.

It caused me considerable anguish because I felt I had failed them. I was certainly not detached enough to say: 'Oh, well, that's business.'

I felt I had a responsibility to look after the livelihoods of everyone who worked for us and I was afraid our sense of community was under threat. Of course, the redundancies weren't a big surprise to much of the workforce and some of their responses were telling – 'You should have done this earlier,' some said – but many of them felt that the board had been negligent in their failure to see what the future held and I couldn't really blame them.

We got a kicking in the media, as usual. I was afraid people might think they were discerning some kind of pattern – falling profits, redundancies, selling part of the operation, when it was all put together it could so easily have been interpreted as a company on the skids, but the more realistic interpretation – and the correct one, as it happens – was that it was a company reinventing itself. And we simply had to do it. We could have stayed as we were and gone down the tubes in 10 years' time, or we could have reinvented ourselves to prepare for the next two decades. It was a messy business, but we knew no other alternative.

THE ISSUES ...

THE DESIGN ISSUE

When we faced up to the reality that we would have to reinvent ourselves, the first thing we had to deal with was the look of the shops. But, looking back, we probably didn't get that right first time. We brought in a design team but they didn't take enough account of our heritage. The Body Shop heritage is visually colourful, full of story-telling and natural ingredients. It should have been simple enough to incorporate all that into a new design, but instead the team came up with something that looked like a clinical laboratory – all glass, plastic, hard-edged aluminium, chrome and very masculine.

One of our problems was that we had too many visual sacred cows and I suppose I was responsible for a lot of them. I wouldn't, for example, hear about ditching dark green as our predominant colour – it is not only symbolic of our involvement in green issues, but it is an important part of the company's heritage and brand trademark.

A word of warning to fellow founder-entrepreneurs – don't let people fiddle and play! Don't let anyone touch or change the DNA of your company unless it is to polish the differences and sing those differences out from every rooftop. And here I speak from experience – the experience of seeing passers-through meddle and shave away years of an earned reputation and then move on.

THE STORY-TELLING ISSUE

Some critics have claimed that our campaigns were turning people away from the shops or that they were getting bored with them. But I knew this wasn't so, because many more people have been signing up for our campaigns than they were in the early days. Four million signed up to register their objections to animal testing and over 12 million made their mark for human rights. The Body Shop is known to be a campaigning shop and people seem to feel comfortable – thrilled, even – to come in and find that the company is doing these things. And they expect them to continue.

But the company has been singularly ineffectual at telling these same people the stories behind the products. We did not master communicating our community trade stories or the work of

The Body Shop Foundation. I think after *Dispatches* we got to be too reluctant to be aspirational. We never quite cracked the problem of communicating our issues and values.

For me, such marketing is about telling stories – where you have come from, what your vision is, the stories behind your products – and I consistently find when I travel, especially to visit pre-industrial groups of people, that their story-telling is the basis of their education.

The Celts had a wonderful definition. They believed that all teachers should be poets, because knowledge is dangerous unless it goes through the heart.

THE FRANCHISE ISSUE

The need to redesign the shops was disguised to some extent by our growth. Over the years operations have expanded enormously – though the company still hasn't yet entered South or Central America, Africa or India – and when you open a shop in a new territory, you do abundantly well because you are so new. What has tended to happen is that our new shops have been doing fine, but the ones that have existed for years have been suffering. This was another issue we had to address.

Running a business in 50 countries around the world and keeping the relationship intimate is very difficult, so that's why the company was reorganized into four regions: the UK, Europe, America and Asia. The head offices remain in Littlehampton and London and continue to be the source of product and marketing ideas, in close co-operation with the regions.

There is a note in our franchise contract that every five years franchisees have to revamp the shops and invest more money into them. Some of them who do so can see the results in increased sales and so are motivated to abide by the contract, but others are 'absentee landlords'. They take their money from the shop and invest in other businesses, restaurants, hotels, whatever. That is their right, of course, but it doesn't do the shops much good. Another problem in Europe is that a number of head franchisees have their own warehouses: there is a warehouse in Switzerland, another in Holland, another in Germany and a fourth in France. It would be much more sensible to centralize distribution, but you've got to persuade the head franchisees, and that's a difficult task.

This kind of conflict of interest is prompting the company to turn away from franchising in favour of partnerships. Franchising was a wonderful way of facilitating world-wide growth, but it doesn't make sense on a world-wide basis any more. It will still be the right way of working in some places, but in others the company began to look instead at partnerships with head franchisees, to give the company more control and the ability to adapt faster.

Turning franchisees into partners meant looking at how we could develop their region together, what we needed from each other, how they could help with new products and who had the best practice. It meant exchanging effort, exchanging employees and exchanging ideas.

THE LESSONS ...

Reorganization was sometimes painful, sometimes liberating, but absolutely necessary – and the challenges we were facing as a retailer were also related to those that every business will have to face up to over the next two decades. In that respect, my own blueprint for The Body Shop could also be a way of looking at how business survives in a changed world. If I had to pick some lessons from recent years to guide us through the future, they would be these:

1. BE QUICK

Speed, agility and responsiveness are the keys to future success. I wanted to be able to act on ideas and discoveries with more speed and get products into the shops much more quickly than ever before. I set up the Creative Greenhouse department, whose job it was to do exactly that. We played with retail concepts, product concepts and design concepts, gave them a gestation period in the stores and then the company took them up if they so wanted. It was almost like a 'skunk-works', a creative group of individuals who were measured by the number of ideas they produced and were not necessarily part of any strategic hierarchy or plan, but more like a genie in a bottle — unsuppressable, with antennae pointed towards areas where the cosmetics industry would never look. It reminds me of a quote I once heard: 'A creative company is no home for peace and serenity.'

2. BE CREATIVE ABOUT DIFFERENT WAYS OF SELLING

One of the things I was shouting about in the company was the need to examine other ways of selling as well as retailing. People are going to go into our shops in the future for an experience and the opportunity to handle the products, but I believed that in the next decade the majority of our customers were going to buy either through home-selling or over the Internet.

After various abortive attempts the company is now experimenting with 'The Body Shop Online' based in the USA. Exactly as predicted sales have been going through the roof. There are no signposts, as far as I'm aware, of a retailing company using not only shops, but also selling on the internet and selling in the home.

I was championing direct selling in The Body Shop long before I knew there was such an expression. I believe it is the future, certainly in this country, as we are now the oldest population in

Western Europe. People like my mum do not perambulate around town, and many people don't live in towns anyway. One of the bleeding needs in our modern society is loneliness, so any of us who has a business that shaves away at loneliness is well placed for the future. And I find no better way than direct selling.

Going into people's homes and ritualising the event is the future. And if you can take with you not only the product, but also the stories behind the product, and the values behind your company, the experience offers the consumer a whole new kind of shopping.

I also believe passionately in the opportunities it provides for women. In every country I've travelled in the West, it is the old larger corporations, dying of boredom, being eroded by giantism, that have lost millions of jobs, and it is the women-owned businesses that have been generating new jobs every year.

I think The Body Shop at Home, which is already operating in the UK, USA, Canada and Australia, will expand enormously and ultimately go worldwide. It is the other end of the spectrum from e-commerce – an opportunity for human contact in a world in which people are going to be increasingly isolated by computers.

3. FORGET IDENTIKIT BRAND

The demise of the retail industry was widely predicted by prophets of the so-called 'New Economy'. But because of our relationship with our franchisees and partners, I don't think the shop-opening rate will slow down over the years to come. On the other hand, I do think there might be different types of stores. You might see a proliferation of just well-being shops, for example.

Shoppers will be looking for something more diverse and I strongly believe the company needs to move away from the idea that every shop has to look the same and that the 'brand' must be protected. Nothing is more boring than going to every town and seeing The Body Shop looking exactly the same. I like the idea that the shops in the Bahamas or Barbados have got the Rasta colours on the walls outside. I like the idea that there is marble inside the shop near the Spanish Steps in Rome.

While there will always be some green in the stores, for example, it is communities that have to be protected, not brands. I would much prefer the shops to have a high profile in the communities in which they trade rather than the brand to be identikit.

4. INTERPRET THE PRODUCT BROADLY

Still, there is so much more that The Body Shop could do that they are not yet doing. I would like to see them moving into education and publishing. The company has 30 years of accumulated wisdom about doing business in a different way and I think they could spread that message. I see them being involved less in fragrance and bubble baths and more in alternative therapies: Ayurveda, homoeopathy, Tibetan medicine and much more.

5. BUILD PARTNERSHIPS WITH COMMUNITIES

Partnership is the key to the relationship with communities and suppliers, because it is a model of new-style business rather than old. The relationship is not a hierarchy, where one gender or rank lords it over another, but a partnership of equals, where both sides can learn from each other. Partnership is also the way The Body Shop organizes their relationships with NGOs as suppliers. The company has about 37 suppliers in 23 different countries and many of those are informal groups or fair trade organizations. The relationship has to be a learning partnership too: each has to learn enough about the other to work with them effectively.

Partnership also covers giving money, which The Body Shop does through their Foundation. Both trading and giving are partnerships. They go where other companies have never dared go before – putting missing persons' pictures on lorries, backing the Ogoni, funding Children on the Edge, which cares for children in Eastern Europe, or Body & Soul, a London-based organization working with HIV+ families.

Emphasis should be on supporting small-scale economic ini-

tiatives wherever the company goes – anything from supporting black family farms in America (almost an endangered species) to building a new fibre economy, supporting new ecological designers or teaching our employees community organizing. This is the way to keep communities vital.

I hope The Body Shop will also be looking at the community of the workplace, making sure it stays open and frank, so that everyone in the organization is capable of being a leader and creativity is never hostage. And remember that even if a highly creative workplace is not always a comfortable place to be, it can be a place where people will argue *gracefully* and where one can be given the space and time to re-examine things.

6. STAY HUMAN AND MEASURE SUCCESS DIFFERENTLY

The Body Shop is Britain's largest international retailer and is probably going to be the world's largest international retailer one way or another at some time, but it has got to retain its humanity at all costs.

It is the values that will keep the company at the forefront in business – being 20 times bigger is not a goal; being better by being values-led is. This means continually auditing the way the business is run and being transparent in everything it does. Building economic growth that respects human rights and sustains communities, cultures, families and the environment. And measuring success accordingly.

I guarantee that in the future the biggest catastrophes in the world will be caused by poverty: poverty of imagination, spiritual poverty and economic poverty. If we carry on the way we are going, what will be common will be forced labour in sweatshops – not just thousands of miles away but also here under our noses. We will see children forced to work long hours, the poisoning of our water and land, dislocation of communities and gross inequalities of wealth. **All this is already upon us and it has to stop.**

In fact, it is probably easier to stop these economic horrors than it is to put a man on the moon – all that limits us is our imagination and our lack of outrage at governments that measure their performance in purely economic terms. Show me one government that measures its greatness by how it treats the weak and frail. I'm convinced that business can show them the way.

7. BE OPEN

My idea is that we need more business regulation, not less. We in business should be penalized when we screw up and make a mess, and rewarded when we can prove that we're heading towards sustainability. I believe every corporation should undergo strict social and environmental auditing.

Regulation leads to innovation and trust. It may involve businesses being more open about their work, but then they should

I have never had a vote, and I have raised hell all over this country. You don't need a vote to raise hell! You need convictions and a voice.

Mother Jones

want to be, simply because transparency is the best way of tackling public cynicism about their motives.

So the true challenge for a business should not be whether it can retain its customers while denying them the information they need. It should be whether it can show customers, employees, suppliers and the world at large an open path from the product on the shelf to the producer in another continent. Business leaders will have to learn that, unless they change their perspective on this, they will suffer lower productivity from disenchanted employees in the short term and desertion by customers in the long term. Accountability is the way forward.

8. MAKE ETHICS PART OF YOUR HERITAGE

We need to develop a corporate code of conduct, a formal, articulated and well-defined set of principles which all global businesses agree to live up to. A broadly kept code of conduct would shut down the excuse about the competition making ethical behaviour impossible once and for all. We must all agree not to compete in ways that destroy communities or the environment. We must all embrace the principles of socially responsible business, because the decisions of business leaders not only affect economies, but *societies*. Unless businesses understand that they have responsibilities they must live up to, in terms of world poverty, the environment and human rights, the future for all of us is pretty bleak. But it has to be more than just a PR whitewash.

9. BE DIFFERENT AND TELL STORIES

Whatever wisdom I have acquired over the years has come from trying to run my business with a difference. We wanted to be different from the very start. We wanted to be utterly honest about the products we sold and the benefits they could offer. We wanted to challenge the status quo in the cosmetics industry. We wanted to incorporate social and environmental change in everything we do, all over the world, every day we are open for business. And I have been unyielding in my commitment to build a business that espouses and practises this new paradigm.

Far too much conventional business thinking is about finding out what your competitors are doing and copying them. At The Body Shop, we looked at the opposition and did exactly the opposite. That meant constantly following unmapped paths, but at least you could feel you were making progress. Probably the most important difference between The Body Shop and conventional businesses is that we raised our sights from uninspiring, short-term ambitions and narrow interpretations of 'progress'.

I have no interest in The Body Shop being the biggest, the most profitable or the largest retailer. I just want The Body Shop to be the best, most breathlessly exciting company – and one that changes the way business is carried out.

10. REMEMBER THAT PEOPLE ASPIRE TO
MORE THAN MONEY

In this global-ecological age, we have got to accept that we are all sisters and brothers, friends and neighbours. And we all need work to be filled with opportunities for personal growth and discovery. It is the foremost job of the new paradigm manager to provide a context in which the spirits of employees can expand and transform.

Spirituality in business is not about esoteric, religious, cosmological or other ephemeral ideas – it is rooted in the concrete action of real people whose sense of care extends beyond themselves. I think that as leaders, we should be making sure that businesses shift their emphasis to an emphasis on the human spirit.

LOOKING TO THE FUTURE

If you look back over the last century, the outlook for founder-entrepreneurs of successful enterprises isn't a very happy one, especially for retailers. Henry Ford flirted with fascism. John Lewis battled with his successors well into old age. Gordon Selfridge lost the plot and spent his days staring up at the store he once owned. William Whiteley was assassinated in his own office. So it's a good idea to reinvent the role every so often.

I am only the founder of The Body Shop, and although I am immensely proud of what we all – staff and customers alike – achieved in the quarter century I spent running it, I know that being the founder does not mean you can or should just "go on and on", as Mrs Thatcher so graphically put it. An organisation of that size has a life of its own and there comes a point where you have to wish it well, and let go.

Of course I have advice. Real success depends so much for the company on what bottom line they live by. Profit is vital, but if that is the only bottom line, any company will shrivel and die. If The Body Shop can stay true to a bottom line that is breath-takingly exciting, empowering and inclusive – not just an Enron-style sleight-of-hand, where all the goodies are for the top managers – then it will stay at the forefront of retailing and business ideas.

This is the modern paradox of business. Sustainable profit doesn't come from an obsession with profit. Neither does change come from an explicit effort to make change, and it absolutely never comes about at the urging of outside consultants or a result of a bloodless strategic planning. But then everyone with any connection with The Body Shop knows what I think.

The Body Shop is itself, and growing in new ways, so it is hardly surprising that it is different now. I follow its development with fascination, of course, and sometimes with frustration – as any founder does – just as I find the development of corporate social responsibility by the rest of the business world so frustrating, in this case for its timidity, slowness and false starts.

When I stepped down as co-chair, I asked myself what role I should play, both in the company and in the world outside, and the answer I gave myself was that I should be an agitator. I wanted to be an irritant, a gadfly – infusing creativity and creating an edge to everything The Body Shop did – and using its values and history as a source of ideas and energy. But I also wanted to do the same independently.

So I have taken that central core of what we did – the activism and communications – and sharpened them so that they have some chance of reaching people and helping them achieve their campaign objectives. **That is what Anita Roddick Publications, AnitaRoddick.com and TakeItPersonally.org are – a way of targeting messages using books, print, electronic media and the internet.**

It is not really intended as another company – I'm not interested in starting another company – but as a method of taking very complex issues, popularising them and changing the world. Part of being effective at anything is being well informed. Any idiot can have an opinion – the people with power to get things done and see their vision realised, are the people who know what they are talking about. That means information really is power. The mission of this communications centre, and by extension my books, is to deliver the best information at the right time and in the right format to the right people, who can use it to effect change in their own lives or in the world. Ideally both. Maybe it doesn't fit into a spreadsheet as easily as units moved, but to me a book sold and a mind opened are, in a way, the same thing. Each of them are an incremental success with the same end purpose.

There are two causes in particular which have been dominating my time. One is the issue of sweatshops and my work with the US-based National Labor Committee (NLC), exposing – for example – the factory in Bangladesh where young women are paid just five cents for every $17.99 Disney shirt they sew, or the Wal-Mart sweatshops in China and Bangladesh. The solution – as so often in the field of corporate responsibility – is enforceable laws to protect human and worker rights such as those currently afforded corporations.

The other has been the campaign for the release of the Angola 3. Albert Woodfox and Herman Wallace have been held in solitary confinement at the Louisiana State Penitentiary at Angola for 31 years for a murder they did not commit. Their friend, Robert Wilkerson, was held in solitary confinement for 29 years, until 2001, when he was released after having proved his innocence in the 1973 killing of another prisoner.

For over three decades, the three men have lived alone in bathroom-sized solitary confinement cells from which they are released only an hour per day. During Louisiana's scorching summer months, the un-air conditioned and barely ventilated cells swelter. They exercise alone, sometimes wearing chains, in outdoor cages about the size of a tennis court, while two of the three living witnesses against them have now recanted and new witnesses have identified another prisoner as the actual murderer.

Woodfox and Wallace were contributors to my book *A Revolution In Kindness*. An advance copy of this book intended for Wallace was confiscated in June 2003 by prison officials on the grounds that it was a 'threat to security'.

Their plight has been behind my donation, together with Gordon, of £1 million to help fund Amnesty International's new Human Rights Action Centre. It is, after all, a critical time for human rights now that the clock seems to have begun spinning backward. We are seeing hard-won rights under attack in the age of the War on Terror, with detentions without trial in Guantanamo Bay, prisoner abuse in Iraq, and high-level intellectual legal arguments about how to get away with torture. I hope the new centre can start pushing the clock forward again.

All these challenges need imaginative communications, and whole new ways of campaigning, and – just like my last decade and a half outlined in this book – there are no signposts that set out how this should be done. Like my last decade or so at The

Body Shop, it feels sometimes like management by falling apart at the seams. Like before, the braver ideas still catch me off guard. Sometimes, I hope like before, I might recapture that feeling of being fraudulent because it is all so easy. But it always feels instinctive, whether I get it right or wrong.

What are the moments in The Body Shop that I was most proud of during the last decade? It was always when the company acted in brave ways. Fighting Shell and the Nigerian dictatorship and then, after five years of campaigning, the release of the Ogoni 19. The human rights campaign, Make Your Mark, where we collected pledges from over 12 million people.

The most painful moments? Having to defend our values in a high profile court case, followed swiftly by a corporate stalker, followed by a huge dent in my reputation when the press decided that I was being shunted aside by the new so-called 'hard-nosed CEO' of The Body Shop. Seeing The Body Shop logo being redesigned into an image I can't relate to and even worse seeing the corporate style and image being less radical. Ah well, I guess everything is subject to change.

Gordon and I both hoped that our daughters would join us in the business. Well, they didn't, though now that they have children they have started to put their toes back in. I'm proud of both my daughters: they have a hugely developed sense of social justice and they never act like rich kids. And they understand and support our decision that they will not be inheriting anything beyond a trust fund and the houses we own. We agree that the function of wealth is not to accumulate it but to give it away as productively and responsibly as you can.

But I'm also proud of my daughters for different reasons. Justine is fearless, and very family-oriented now she has Maiya and Atticus-Finch. Sam is one of the best-defined activists I've ever come across, though she's shifting her outrage now that she is the mother of O'sha-Bluebell. I am looking forward to the fun to be had from this gaggle of grandkids in my life.

There are bad days when I'm consistently depressed about the perils of globalization. The world certainly looks bleaker to me now than it did five years ago. But on a good day, I remember that businesses are not dropped from the skies ready-formed; they are created and shaped by human beings and they can be changed by them if they have the will to do it.

Never be seduced into believing it isn't the role of business to tackle the big issues, because it absolutely is.

Walt Whitman once wrote:

This is what you shall do: love the earth and sun and the animals, despise riches, give alms to everyone that asks, stand up for the stupid and crazy, devote your income and labor to others, hate tyrants, argue not concerning God, have patience and indulgence towards the people, take off your hat to nothing known or unknown, or to any man or number of men, go freely with powerful uneducated persons, and with the young, and with the mothers of families, re-examine all you have been told in school or church or in any book, and dismiss whatever insults your own soul; and your very flesh shall be a great poem, and have the richest fluency, not only in its words but in the silent lines of its lips and face, and between the lashes of your eyes, and in every motion and joint of your body.

I took these words to heart years ago, because Whitman delivered the message about the real beauty of living much better than I ever could. When I look back, I can see how his What eventually became my Why, the inspiration for the being of The Body Shop. And reading his words again, I can also see how far I still have to go.

INDEX

The job of a citizen is to keep his mouth open. Gunter Grass

Troubled Water

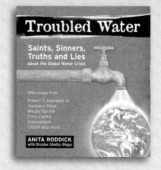

ANITA RODDICK
with Brooke Shelby Biggs

Worldwide, a billion people lack access to clean water. Droughts, floods, and waterborne diseases kill millions every year. Multinational corporations see a profit opportunity unparalleled by oil or even gold, and are buying up and selling a basic human need. Meanwhile, we consume millions of bottles of designer water every day. Why are the politics of water so skewed, and what can we do about it? This book provides elegant answers to these hard questions.

ISBN 0-9543959-3-X
Retail Price: UK £9.99   US $17.95 Paperback
(Volume, non-profit, and educational discounts available. Please call for pricing.)
From US telephones: 1-800-639-4099 From UK telephones: 0800 018 5450
Or visit www.anitaroddick.com, www.troubledwater.org or www.takeitpersonally.org

BRAVE HEARTS REBEL SPIRITS

WRITTEN BY BROOKE SHELBY BIGGS.
CONCEIVED BY ANITA RODDICK

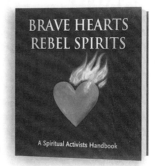

You know the names Martin Luther King, Jr., Mohandas Gandhi, Nelson Mandela. But have you heard of Roy Bourgeois, Neta Golan, or Sulak Sivaraksa? They, and the dozens more spiritual activists in this book are the heirs to that great tradition of faith-based activism.

The spiritual activists in this powerful, provocative and visually stunning book are environmentalists and gay-rights activists, peace workers, land reformers and child advocates. They are Buddhists and Catholics, Hindus and Muslims, Baha'is, Jews and Quakers. The stories of these modern-day prophets of positive change will inspire you, and the resources provided in each chapter will help you put your own beliefs to work in the world.

ISBN: 0-9543959-0-5
Retail Price: UK £12.99, US $18.95
(Volume, non-profit, and educational discounts available. Please call for pricing.)
From US telephones: 1-800-639-4099 From UK telephones: 0800 018 5450
Or visit www.anitaroddick.com